Calculation of Seismic Actions for Structural Analysis

Enrico Zacchei · Reyolando M. L. R. F. Brasil

Calculation of Seismic Actions for Structural Analysis

Methodological Approaches and Experiences

 Springer

Enrico Zacchei [ID]
Salamanca, Spain

Reyolando M. L. R. F. Brasil
Barueri, São Paulo, Brazil

ISBN 978-3-032-08883-3   ISBN 978-3-032-08884-0   (eBook)
https://doi.org/10.1007/978-3-032-08884-0

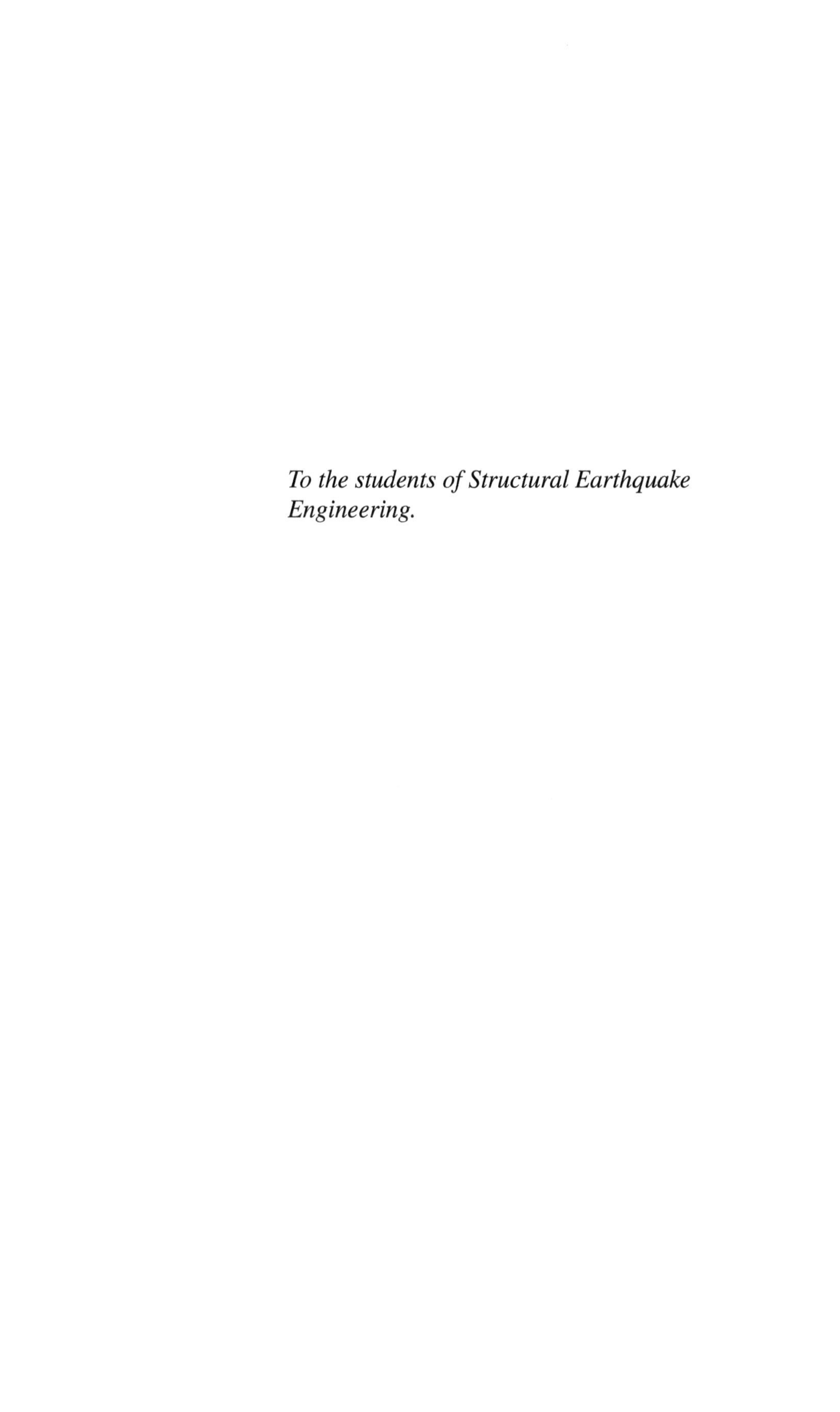

To the students of Structural Earthquake Engineering.

Foreword

In earthquake engineering, seismic input refers to the quantification of the ground motion that a structure is expected to experience during an earthquake. Accurately defining this input is essential for assessing structural response, ensuring safety, and guiding endurable structural design. Despite its fundamental role, this aspect of structural analysis is often treated briefly in engineering education and literature, due in part to its inherent complexity and the interdisciplinary knowledge it demands.

Calculation of Seismic Actions for Structural Analysis—Methodological Approaches and Experiences addresses this issue with clarity, rigour, and a strong pedagogical focus. It offers a comprehensive exploration of the methodologies used to calculate seismic input, framed within a broader conceptual and methodological discussion that makes the material accessible to readers with diverse academic backgrounds. The authors succeed in making sophisticated material accessible without sacrificing depth. Numerous diagrams, equations, and explanatory notes support the text, offering clarity and reinforcing understanding.

The book introduces eight complementary approaches, ranging from simplified and direct methods to advanced procedures such as probabilistic seismic hazard analysis and physically based models. Special attention is paid to fundamental concepts of seismology, the mechanics of seismic waves, and the modelling of soil-structure interaction-topics essential to understand and apply seismic input methods effectively.

A distinctive feature of this work is its balance between theory and practice. Five detailed case studies, based on real research and engineering projects, demonstrate the application of these methodologies to various structural systems, including dams, industrial facilities, and civil infrastructure. These examples greatly enhance the book's practical value, bridging the gap between academic knowledge and real-world engineering challenges.

Dr. Enrico Zacchei brings to the book a rare combination of academic expertise and international professional experience. His background spans several countries, regulatory contexts, and project types, offering the reader a global and adaptable perspective on seismic design. Prof. Dr. Reyolando M. L. R. F. Brasil contributes decades of excellence in structural and geotechnical engineering education and research, further reinforcing the book's academic depth.

This volume is intended for advanced undergraduate and graduate students in civil and structural engineering, as well as professionals and researchers involved in seismic design. It may serve as both a textbook and a reference for professionals seeking to deepen their understanding of seismic action modelling across a variety of structural and geotechnical contexts.

The clarity of exposition, methodological depth, and practical insight offered throughout make this a valuable and timely contribution to the literature on earthquake engineering.

Pablo Moreno
Full Professor
Department of Mechanical Engineering
University of Salamanca (USAL)
Salamanca, Spain

Preface

The purpose of this textbook is to summarize in a didactic way several methodologies for calculating seismic input in the design of buildings, civil and industrial structures. Rather than focusing solely on mathematical equations (which are there anyway), the aim is to provide a clear explanation of the parameters involved in these methodologies. The book adopts a didactic and methodological perspective, ensuring that both basic and more complex concepts are presented in a way that allows readers to grasp the underlying general principles. It is designed to facilitate a deeper understanding of how seismic design calculations are approached, helping mainly undergraduate and graduate students apply these methods effectively.

The first author of this book, Enrico Zacchei, has been dealing with earthquakes directly or indirectly for a long time; from 2009, when he was finishing his engineering studies, placing a particular emphasis on earthquake engineering (exactly when L'Aquila earthquake occurred), to 2023, when he was analysing embankment dams under earthquakes (by the time Turkey-Syria earthquake occurred). The second author, Reyolando Manoel Lopes Rebello da Fonseca Brasil, a specialist in earthquake engineering, supervised and strongly supported the entire work.

The main concept in designing safe, resilient and sustainable structures is correlated to how a methodology must be applied, when it must be applied and why it must be applied. The latter question is strongly correlated to a cultural problem regarding prevention. Prevention is the best form of action for most fields!

The cultural problem is longstanding. In the West in 1703, when a violent earthquake occurred in the Marche region in Italy, the salvation of a nearby city was attributed to Saint Emygdius. In the East, around the 16th century the people in the area around Lake Biwa, in Japan, associated the causes of the earthquake to the movement of a giant underground catfish called "Namazu". At least the Japanese understood that an earthquake came from the underground and not from the sky. Who knows if this influenced the advanced of seismic risk studies? In Japan compared to Italy they were certainly more attentive to the problem.

Book Outline

The calculation of seismic actions is a fundamental aspect of structural engineering, particularly for buildings and civil or industrial structures located in or near seismically active areas. A comprehensive understanding involves learning how to predict, analyse, and mitigate the effects of seismic forces on structures.

This is a conceptual and didactic book on this topic, where eight methodologies to define the seismic action and five case studies of structure under earthquakes are shown.

In Chap. 1, the general concept of seismology and earthquake (i.e., seismic waves, ground response and magnitude) has been introduced. From Chaps. 2 to 6, five main methodologies have been explained in sequence: elastic response spectrum, artificial accelerogram, probabilistic and determinist seismic hazard analysis, and physically based seismic hazard analysis. Also, three simplified and direct methods are explained (Chap. 3). In Chap. 7, five case studies (technical and research projects) where the authors have participated in the analyses and design are shown. Finally, in Chap. 8, for completeness three appendices (A, B, and C) are listed. Several figures and tables help to explain these methodologies and case studies.

<table>
<tr><td>Salamanca, Spain</td><td>Enrico Zacchei</td></tr>
<tr><td>Barueri, Brazil</td><td>Reyolando M. L. R. F. Brasil</td></tr>
</table>

Acknowledgments

Without the contributions of numerous professionals, this book would not have been possible. It brings together a wealth of experiences, materials, and valuable advice.

For this, authors thank Prof. Fabio Sabetta for the didactic material to develop the methodologies in Chap. 5, provided during his seismology course; Prof. Lawrence Hutchings for his precious comments and suggestions for understanding the methodology in Chap. 6; Prof. Fernando Stucchi, director of the Brazilian company EGT Engineering and Prof. Pedro Lyra for the technical support to carry out the analyses in the Case study 1, and Prof. Fabrizio Paolacci to direct the first author's master's degree regarding this case study; Prof. Antonio Tadeu, director of Portuguese institute ITECONS, for technical meetings to develop the Case study 2; the colleagues of the Italian working group ITCOLD to develop the Case study 3; Prof. José Luis Molina to direct my Ph.D. Thesis on the Case study 4; and Eng. Augusto Reis Junior, for the precious advice on the industrial plants (Case study 5).

Despite all comments, suggestions, discussions received from these colleagues, all considerations in this book are to be attributed only to the authors of this book.

General Notes

The authors have attempted to use the same nomenclature for parameters and coefficients throughout the book. Most methodologies are explained using different national and international codes, books, papers, etc.; thus the used nomenclature may differ from each other. However, for each chapter the same nomenclature is strictly respected.

Conceptually in the use of all codes and literature, the theories and approaches are the same and the authors have always been careful to explain it to avoid confusion. In this sense, in this book, the authors explain the procedures so the readers can understand the physical meaning and the general concept.

All figures have been adapted using the sentence "adapted from," and referenced accordingly. Some figures have been retrieved from iStock (https://www.istockphoto. com/es) and other websites, as indicated in each figure. Other figures were elaborated by the authors.

The used database (all of them are free) has been indicated, and scientific papers, technical papers, etc., have been downloaded by using the rights (when applicable) paid by the University of Salamanca (USAL), Spain, and the University of São Paulo (USP), Brazil, where the authors work.

The used software has been referenced, and the licences have been paid when the analyses (technical and scientific) have been carried out. The licences of Wolfram Mathematica (version 12.0, Wolfram Research, Inc., 2019) and AutoCAD (version 2010, Autodesk, Inc., 2010) were provided by the University of Salamanca (USAL), Spain. The licence of Sap2000 (version 16.0 Plus. Computers and structures, Inc., California/New York, 2013) was provided by Brazilian company EGT Engineering. The licence of SCIA Engineer (version 2019) was provided by ITECONS institute, Portugal. The Deepsoil and Crisis (version 7.0) have a free licence.

Contents

About the Authors

Enrico Zacchei is Senior Structural Engineer, Researcher, and Adjunct Professor. He holds a Ph.D. from the University of Salamanca, Spain, completed in co-tutorship with the Polytechnic School of the University of São Paulo, Brazil, and a second Ph.D. from São Paulo State University, Brazil. He also earned both his Master's and Bachelor's degrees in Civil Engineering-Structures from Roma Tre University, Italy. With over 15 years of technical and research experience in leading national and international entities in Italy, Brazil, and Portugal, he has been involved in more than 20 projects. His academic output includes over 45 papers published in high-impact journals (JCR, Scopus) and more than 25 conference presentations, many of them as a guest speaker. He has carried out numerous projects and publications regarding earthquake engineering. He has undertaken research visits and teaching collaborations in Belgium, France, Poland, and India. He is active member of several professional and scientific organizations, including ITCOLD-Italy, SPES-Portugal, RENIDES-Chile, and MDPI-Switzerland, and serves on several technical and scientific conference committees.

Reyolando M. L. R. F. Brasil holds a degree in Civil Engineering from the Mackenzie School of Engineering, Brazil, and has always worked professionally in structural consulting. He holds a Master's degree, a Ph.D. and a Full Professorship from the Department of Structural and Geotechnical Engineering at the Polytechnic School of the University of São Paulo (USP), Brazil. He was Full Professor of Materials Strength at the Mackenzie School of Engineering and Associate Professor in the Department of Structural and Geotechnical Engineering at the Polytechnic School of USP and is currently Full Professor of Aerospace Structures at the Federal University of ABC, Brazil. With 50 years as Professor and Researcher, he has published several books, over 100 articles in international journals, and has supervised 10+ doctorates and 40+ master's degrees. He is CNPq Research Productivity Scholar, level 1B.

Chapter 1
Seismology and Earthquakes

Abstract This chapter introduces the fundamental concepts and parameters of earthquakes, such as seismic waves and magnitude. These provide the foundation for understanding the approaches used to define seismic actions throughout this book. The physical and mathematical description of body and surface wave propagation forms the basis for seismic analyses of soil and structures, while magnitude serves as the most universal and widely recognized measure of an earthquake.

1.1 General

The study of earthquakes involves various fields and specialists, such as geophysicists, geotechnical engineers, and structural engineers. Each specialist approaches the phenomenon differently, by analysing the mechanisms that generate the deep fractures in the Earth where an earthquake occurs and its energy is released, examining the movement of various soil layers due to the passage of the seismic waves, and studying the resulting oscillations of structures.

The perception of an earthquake by humans and animals is basically the shaking of the earth, and for this reason, since more than 2000 years ago, the majority of approaches have been based on recordings of the ground through a simple pendulum, which measures the oscillations of solid bodies. Lucretius described two famous earthquakes of antiquity, that of "Sidone" in Syria (400 BC) and that of "Egio" in the Peloponnese (372 BC), which he perceived as a shaking of the earth, mistakenly associating them with "subterranean winds" [2].

Shaking is described by the movements of the ground that in turn move the structures and infrastructures (and in general all solid bodies) that are placed on the ground. The surface structures can move differently with respect to the ground and in different directions, causing situations that are difficult to predict, in terms of individual responses of the soil and the structure, and in terms of interactions between them. For all cases the structures also have structural elements within the ground itself (i.e., shallow or deep foundations) that can therefore move "as a whole".

E. Zacchei and R. M.L.R.F. Brasil, *Calculation of Seismic Actions for Structural Analysis*, https://doi.org/10.1007/978-3-032-08884-0_1

The movement is described physically by the displacements of the ground mainly in the vertical and horizontal directions, however it is common practice in engineering to treat and model these movements using accelerations, which are the second derivative of the displacements in time. The reason is that in this way it is possible to estimate the inertial forces acting on the structure that depend on the masses of the structural elements (easy to calculate). In fact, buildings and civil structures have large masses so the inertial forces represent important forces to be considered. Furthermore, working in terms of displacements could be more complicated (at least analytically) because it would be necessary to work by using the compatibility equations instead of the static equilibrium equations.

The goal of the structural engineers is to safeguard the structures built on the earth's surface and the people inside them from an earthquake. A good "metaphor", described in [1], which gives an idea of the vulnerability of our structures, is that the Earth and its crust is like a whole egg with a cracked shell where our structures are placed.

Since about the beginning of the 1970s, recording media have moved from an analogue system to a digital system greatly improving the quality of the signal and capturing all the information of the earthquake. Figure 1.1 shows an example of a seismogram where a pendulum with a "stylus" writes the ground accelerations over time on a roll of paper.

Some main aspects to characterize an earthquake are the focal mechanisms (or fault ruptures), and the fault-site distances. The focal mechanism indicates the type of rupture between the rock blocks that produces the earthquakes. The focal mechanism is identified by its mechanical behaviour and its rupture surface called "seismic fault".

There are three basic types of the focal mechanism as shown in Fig. 1.2:

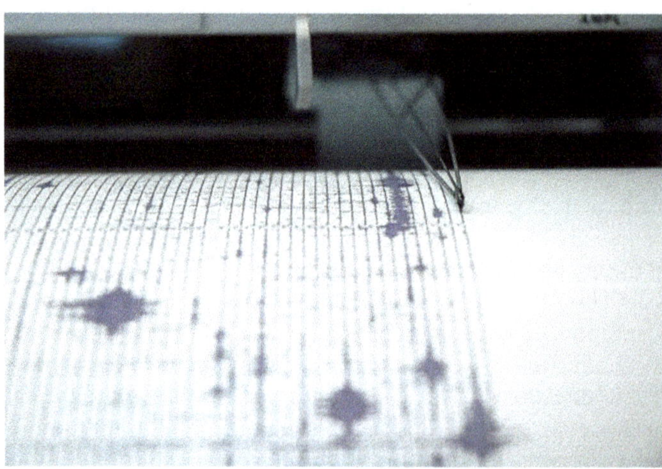

Fig. 1.1 Example of an earthquake registration by a seismogram (by iStock site)

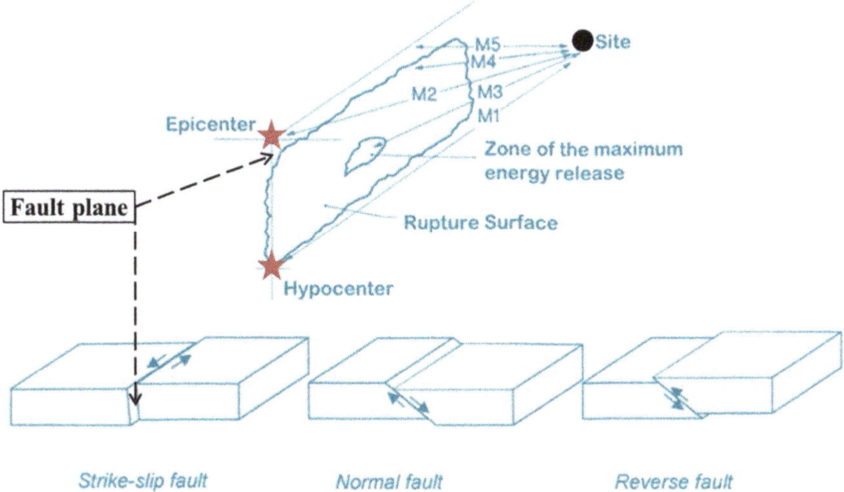

Fig. 1.2 Scheme of the fault-site distances (above) and focal mechanisms (below) (adapted from [11])

  (i) Strike-slip fault. It consists in a rupture surface developed in a vertical fault plane where a relative horizontal displacement of the two blocks occurs. The horizontal movement can occur either to the right or to the left. This displacement generates the shear stresses in the crust that act tangential to the surface.

 (ii) Normal fault. It occurs when the horizontal component of dip slip movement is extensional and when a material block moves downward relative to the material above the fault. The fault plane in inclined. This mechanical movement is associated with tensile stresses in the crust.

(iii) Reverse fault. It like the normal fault, but the dip slip movement is compressional, and the material moves upward relative to the other material block. Here the mechanical mechanism is characterized by compressive stresses, thus results in a horizontal shortening of the crust. When a reverse fault has a small dip angle with respect a horizontal line is called "thrust fault" that can produce very large movements generating great earthquakes, this because the total force resultant is very similar to the horizontal force component. respect a horizontal line is called "thrust fault" that can produce very large movements generating great earthquakes, this because the total force resultant is very similar to the horizontal force component.

These three mechanisms represent the fundamental ones in the mechanics of materials and structures (i.e., shear, tension, and compression). They can be combined with each other providing more complete mechanisms; this schematic division is only useful to identify the predominant focal mechanism to characterize the seismic zone. However, in real situations they are more complex, and irregular as shown in Fig. 1.3

Fig. 1.3 Real seismic fault (by iStock site)

where it is also possible to see that the irregular fractures have reached the superficial ground.

The fault-site distance (or source-to-site distance) is defined as the reference distance between the observation point (site) where it is necessary to know the seismic input. The observation point can correspond to a seismic station and/or to a structure to be built or already existing.

There are different source-to-site distances, such as (see Fig. 1.2):

(i) Hypocentral distance (named "M1"). It is the distance from the earthquake hypocentre (i.e., the point in the Earth where the rupture starts) to the site. The hypocentre should represent the origin point of an earthquake underground (focus of the earthquake), however given that the fault is an irregular area the identification of a unique point is purely a practical simplification.

(ii) Epicentral distance (M2). It is the distance from the earthquake epicentre to the site. The epicentre is simply the vertical projection of the hypocentre, thus the difference between both points represents the depth of the earthquake. This distance is subjected to the same positional approximations of the hypocentre.

(iii) Distance from the most energetic zone (M3). It is the distance from the zone of maximum energy release to the site. This distance could be the most interesting and useful since it refers to the earthquake energy. As discussed later, the energy in general is the best parameter to quantify both an earthquake and mechanical behaviour of a structure.

(iv) Distance from the fault (M4). This is the minimum distance between the site and the fault. Also, in this case it is difficult to select a unique point in the fault. This distance could be quite conservative since the use of a short distance increases the seismic input (well discussed in Chap. 5).

(v) Distance from the surface projection of the fault (M5, also called "Joyner-Boore distance"). This is the minimum distance between the site and the fault projection on the ground surface. If the superficial fault is visible (as shown in Fig. 1.3) it is not difficult to be estimated. It indicates a superficial distance like the epicentral distance.

It is evident as there is no unique way to define a distance and measure it, indicating the difficult to quantify the earthquake size in exact way, in particular the ground motion, since the distance is a key parameter to define it in analytical and numerical way. Also, the seismic source, here represented by a rectangular area, could be modelled by a simple line (e.g., linear fault), point (e.g., hypocentre), circular area, etc., and then they can be projected into the ground. This mainly regards the necessity to model in mathematical way the physical phenomenon.

The estimation and choice of a distance also depends on the considered seismic waves, which can assume different velocity thus they arrive in different time in a network of far/near field seismic stations. These aspects will be discussed in the following sections.

1.2 Wave Propagation

1.2.1 Body Waves

Consider an infinite rod (or unbounded media) where a one-dimensional seismic wave propagates in the parallel direction to this rod, called x-direction, as shown in Fig. 1.4. The presence of the rollers (idealized mechanical supports) in the superior and inferior face of the rod avoids radial straining.

Consider also that the linear elastic vibrations are free and parallel to the x axis. Under a uniform axial stress that causes an axial deformation, the displacement in time, t, of the material particle, u, is [3, 5]:

$$\frac{\partial^2 u}{\partial t^2} = v_p^2 \frac{\partial^2 u}{\partial x^2}, \quad v_p = \sqrt{\frac{G(2-2v)}{\rho(1-2v)}} \tag{1.1}$$

where v_p is the propagation velocity of the longitudinal primary wave, called p-wave. This velocity, v_p, can calculated in function of the shear modulus, G, Poisson's ratio,

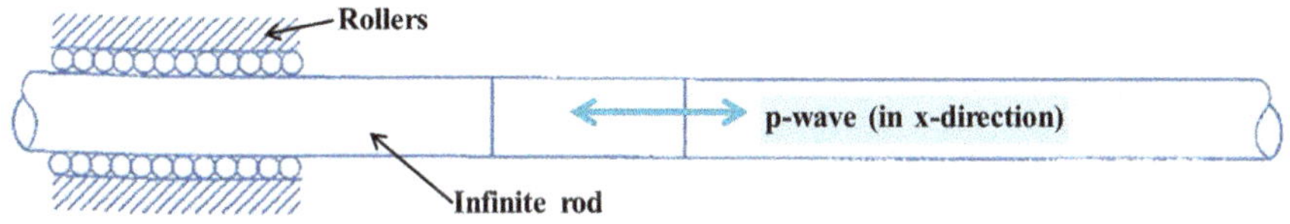

Fig. 1.4 Infinite rod subjected to a longitudinal wave (adapted from [3])

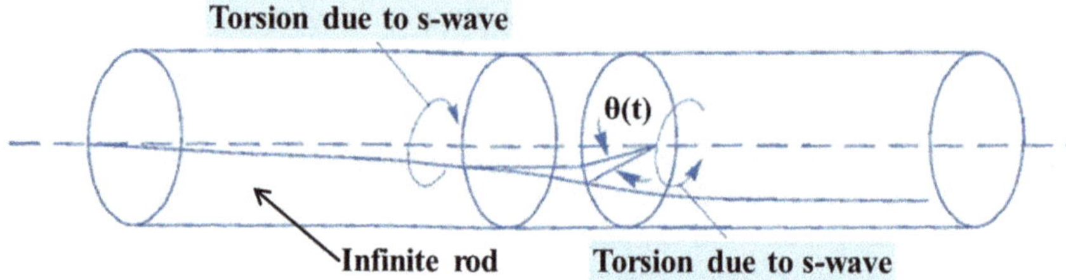

Fig. 1.5 Infinite rod subjected to a shear wave (adapted from [3])

ν, and density, ρ, of the material of the rod, in accordance with the known Hooke's law for an isotropic, homogeneous and linear-elastic material.

An isotropic body indicates that its properties are the same in all directions, whereas the homogeneity indicates that the material is made of the same composition. The linear-elastic behaviour regards the direct correlation between the external forces applied on the medium with its internal deformations, which are linear and do not exceed a certain limit (for further details, it is recommended to read classical books as for instance [12]).

For the upper limit of the Poisson's ratio, i.e., $\nu = 0.50$, the velocity $v_p \to \infty$ indicating that the body becomes incompressible, i.e., infinitely stiff with respect to dilatation deformation. This velocity of the wave, v_p, is not the same as the velocity of the material particle, which is defined as $\dot{u}(t) = \partial u / \partial t$. The particle refers to the medium where an external seismic wave passes through it.

Consider now a torsion of the mentioned rod that causes a shear strain as shown in Fig. 1.5. The cross-section planes remain planar during this torsion avoiding variations of the rod length and alterations of the shear stresses.

The angle of the torsion, $\theta(t)$, of the material particle is:

$$\frac{\partial^2 \theta}{\partial t^2} = v_s^2 \frac{\partial^2 \theta}{\partial x^2}, \quad v_s = \sqrt{\frac{G}{\rho}} \tag{1.2}$$

where v_s is the propagation velocity of the shear secondary wave, called s-wave (the x-axis is parallel to the horizontal dashed central line). Note that both wave velocities (v_p and v_s), only depends on the stiffness and density of the soil.

By considering a steady-state harmonic deformation described by a $\cos(\cdot)$ function, the solution of the partial differential Eq. (1.1) can be written as:

$$u(x, t) = A \cos(\overline{\omega} t - kx) + B \cos(\overline{\omega} t + kx) \tag{1.3}$$

where A and B are the amplitudes of the displacement, $u(x, t)$, to be estimated by defining the appropriate boundary conditions, $\overline{\omega}$ is the circular frequency of the wave, and k is the wave number estimated by $k = \overline{\omega}/v_p$ (or $k = \overline{\omega}/v_s$).

The wave number, k, is correlated to the wavelength, λ ($= 2\pi/k$), which, in a x-domain, represents the distance between two following peaks (see Fig. 1.6). In a

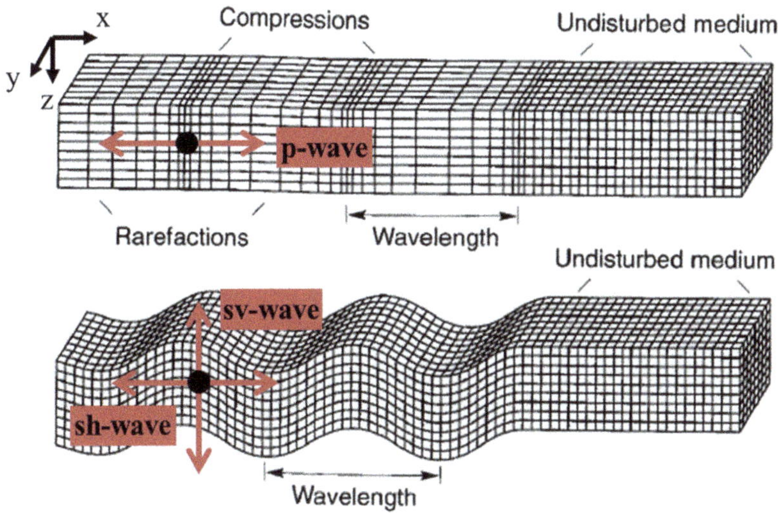

Fig. 1.6 Deformation of the elastic medium produced by body compressive, p (above), and shear, s (below), waves (adapted from [3])

t-domain the distance between two following peaks is defined by wave period, $\overline{T}$ (= $1/\overline{f}$, where $\overline{f} = \overline{\omega}/2\pi$, is the wave frequency).

Note that Eq. (1.3) has a positive and negative part, since the hypothesis is that the wave, which is generated in a specific point, travels at the same speed, v (v_p or v_s), in the rod in both horizontal directions to the right and to the left. In nature, the waves spread from a seismic source in all directions (x, y, z) up to the surface. Both described waves are idealized as waves that propagate in only one dimension.

By using the complex notation and only the positive x-direction (i.e., B = 0), the solution could be:

$$u(x, t) = Ce^{i(\overline{\omega}t - kx)} \tag{1.4}$$

where i is the imaginary part, and C is the amplitude of the wave displacement (C $\neq$ A, since the function form is different, however both amplitudes give the physical meaning to the displacement).

The mathematical form of Eq. (1.4) can be verified by substituting it in Eq. (1.1), however it not the unique solution since other forms could be valid. In an analogue way, it is possible to define a solution of Eq. (1.2).

Figure 1.6 shows the deformation of the elastic medium (elastic soil) produced by the passage of the body waves (p-wave, s-wave). The black point represents the particle of the medium excited by the waves. Here it appears more clearly as the elastic medium deforms parallel to the vibration of the longitudinal p-wave (compressions) without rotations (irrotational wave), and in other parts of the elastic medium the soil is rarefied (rarefactions); whereas it deforms transversally under the shear s-wave without volume change (equi-voluminal). In the parts far from the wave the medium is undisturbed.

Due to the distortion of the elastic medium under the passage of the s-wave, two components are created, i.e., the horizontal (sh-wave) and vertical (sv-wave) component [4]. Thus, the motion of the s-wave is represented by the vector sum of these two components. Note that the s-wave changes its direction during the path,[1] and when it arrives on the surface the sv-wave component becomes horizontal with respect to the surface. For this reason, it is the most important wave to be considered for buildings and structures in general since it moves the ground horizontally.

It is intuitive to see that the p-waves are faster than the s-waves. This is because the longitudinal vibrations arrive first, however they depend on the material characteristics as already mentioned. For a Poisson's ratio of $\upsilon = 0.30$, $v_p = 1.87v_s$. In a real rock, $v_p \approx 1.60v_s$ (with an order of magnitude of $v_s > 800.0$ m/s), and in a soft soil $v_p \approx 12.50v_s$ (with an order of magnitude of $v_s < 180.0$ m/s). The values of v_s tend to zero (i.e., $v_s \to 0$) in an inviscid fluid, which provides no resistance to the shear stresses. This is another reason that helps to understand the fact that s-waves are called secondary waves.

1.2.2　Surface Waves

When the seismic body waves arrive near to the surface, the combinations of the mentioned components generate other type of waves, called surface Rayleigh and Love waves.

The surface Rayleigh waves are formed by combining the p-waves and s-waves. Given that s-waves have two components, all three components generate a retrograde elliptical movement to the material particle (as mentioned, by the Snell's law the sv and sh components reverse near to surface, Fig. 1.7 above). These effects are described by dilatation and rotation functions.

Consider a semi-infinite body with a planar free surface. This free surface is a plane described by the x- and y-axis, where x-axis is parallel to the wave direction, whereas the z-axis is the depth. The displacements due to the passage of the Rayleigh wave in x- and z-directions, during the time, t, are defined as, respectively:

$$u(x, z, t) = -A_1 i k_R e^{-qz+i(\overline{\omega}t-k_R x)} - A_2 s e^{-sz+i(\overline{\omega}t-k_R x)} \tag{1.5}$$

$$w(x, z, t) = -A_1 i k_R e^{-qz+i(\overline{\omega}t-k_R x)} + A_2 i k_R e^{-sz+i(\overline{\omega}t-k_R x)} \tag{1.6}$$

---

[1] This phenomenon is described by the Snell's law, which estimates the direction change of ray paths at interface between different materials. It is a geometrical law that quantifies the angle of reflection (waves that come back) and the angle of refraction (waves that advance up to the surface) of the incident wave.

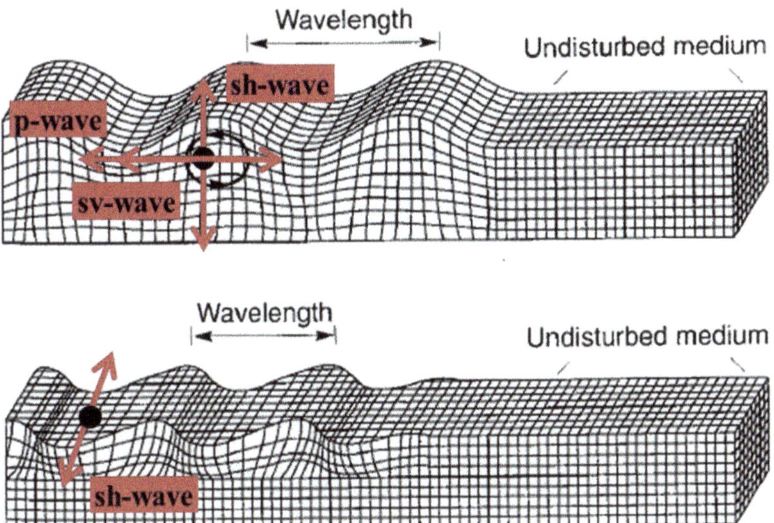

Fig. 1.7 Deformation of the elastic medium produced by surface Rayleigh (above) and Love (below) waves (adapted from [3])

where A_1 and A_2 are the amplitudes of the displacements, and k_R is the Rayleigh wave number ($k_R = \overline{\omega}/v_R$, where v_R is the Rayleigh wave velocity). The factor q and s depend on v_p and v_s, respectively, indicating their direct correlation between the Rayleigh waves with the p- and s-waves. The parameters v_p, v_s, $\overline{\omega}$, i, have been already explained.

The horizontal, u, and vertical, w, displacements are out of phase by 90°, thus, the w displacements reach maximum values for u = 0. The u displacements involve a greater superficial layer ($z \approx 0.50\ \lambda_R = 0.50\ 2\pi/k_R$) with higher amplitudes than the w displacements. For deep layers the horizontal displacements tend to rapidly assume very low amplitudes.

The other surface waves are the Love waves, which substantially consist of sh-waves that vibrate transversally to the wave direction. The horizontal displacements of the soil in y-direction, during the time, t, is defined by:

$$v(x, z, t) = B_1 e^{-v_1 z + i(k_L x - \overline{\omega}t)} + B_2 e^{v_1 z + i(k_L x - \overline{\omega}t)} \tag{1.7}$$

where B_1 and B_2 are the amplitudes of the displacements, and k_L is the Love wave number ($k_L = \overline{\omega}/v_L$, where v_L is the Love wave velocity, which has the same order of magnitude of v_s). In an analogue way as described by Eq. (1.3) all these amplitudes can be obtained by appropriate boundary conditions.

The Love waves are trapped by multiple reflections within the superficial layer, and their amplitudes depend on the velocity of downgoing, v_1 (superior layer 1), and upgoing, v_2 (inferior layer 2), waves. Considering the first layer H, where usually the amplitudes of the Love waves are large, Eq. (1.7) is valid for $0 \leq z \leq H$, where B_1

and B_2 depend only on v_1 velocity. For $z > H$, the horizontal displacements depend on v_2 velocity, and they assume very low values.

Figure 1.7 shows the deformation of the elastic medium produced by the surface waves. The black point is the material particle subjected to the waves.

These surface waves are very important, in particular the Rayleigh waves since they have similar velocity to the s-waves (i.e., $v_R \approx v_s$, for $0 < v < 0.50$) and maintain high amplitude for far distance. Instead, the Love waves appear more dispersive.[2]

Other types of waves exit but they are much less significant from a structural engineering standpoint. Some examples can be (i) the tertiary t-wave, which are quite slow, and they are usually registered in sea water, and (ii) the Lg-wave, which is formed by all the rays of the Rayleigh and Love waves, with a frequency between 0.30 and 3.0 Hz [7].

1.2.3   Ground Response Analysis

The one-dimensional model already described for body waves, in particular for the s-waves, allows to define the analytical models to estimate the ideal ground response. The analytical models can be divided in four different types in function of the behaviour of the soil and bedrock: (i) uniform undamped soil on rigid rock; (ii) uniform undamped soil on elastic rock; (iii) uniform, damped soil on rigid rock; and (iv) uniform, damped soil on elastic rock.

The main differences between these models regard the dissipation capacity of the superior soil (thus its capacity to reduce the wave amplifications before reaching a structure) and the deformability of the inferior bedrock (thus its interaction with the superior soil).

The general concept of the model, shown in Fig. 1.8, is that the registration of a seismic acceleration in the time-domain, t, in a point R is the same to that in the point A, since a bedrock does not alter the acceleration values for relatively high distances. This would allow to know the accelerations of the deep bedrock (R point) that cannot be directly registered. Thus, by using the Fourier spectrum and the transfer function, both in the frequency-domain, $\overline{\omega}$, it is possible to estimate the acceleration in the point S (vertical projection of R point).

The transfer function is a complex function defined as the ratio between the soil response at point S with respect point A. The response can be expressed in terms of displacements, accelerations, etc. In this way, it is possible to estimate

---

[2] The dispersion, the scattering effects, as well as the material damping, radiation damping, etc. are important phenomena that alter the displacements of the soil in surface. They are not treated here; however, in general, the dispersion and scattering are correlated to the wavelengths, whereas the radiation damping is correlated to the path distance. The material damping (the most important effect) regards the non-linear behaviour of the material, and it is correlated to its energy dissipation capacity.

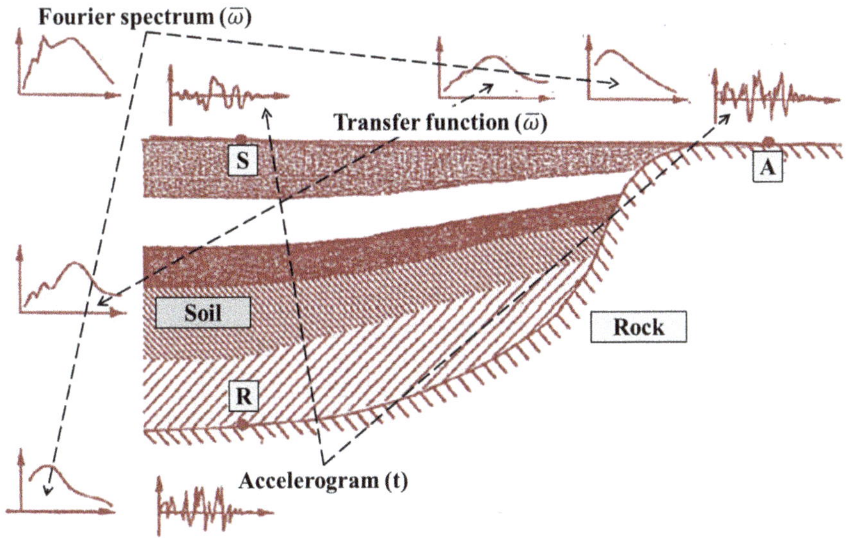

Fig. 1.8 Ground response: general concept (adapted from [5])

the amplifications or damping[3] of the seismic waves due to the characteristics of the different soil layers (for more details see [3, 5]). The Fourier spectrum will be discussed later.

The first model "uniform undamped soil on rigid rock" is the simplest with respect other ones, however, its approach and methodology allows to solve all four models. The last model "uniform, damped soil on elastic rock" appears the more like a real soil and for this it is explained as follows.

The hypotheses of these analytical models are the same defined for the wave propagation model, i.e., the layers are isotropic, homogeneous and linear elastic, to be able to use the previous relations. The consider soil (here subscript "s") of height H rests on a half-space of bedrock (subscript "r"). The soil and rock layers are assumed to extend infinitely in the horizontal direction. They are excited by a harmonic s-wave that propagates in the vertical direction, z (z_s and z_r), and incident in perpendicular way to the rock and soil surface.

In particular, Fig. 1.9 shows the uniform, damped soil on elastic rock model, under a one-dimensional seismic wave. As mentioned, the s-wave has two components (sh- and sv-wave) where it can be assumed that the predominant motion is provided mainly by the sv-wave.

The horizontal motion in time, t, of the soil, $u_s(z_s, t)$, is analytically described by [76]:

---

[3] The amplifications and/or damping of the soil response can be quantified by the resonance phenomenon (i.e., $\bar{\omega} \approx \omega_1$, where ω_1 is the fundamental circular frequency of the soil) and by the impedance ratio, I, calculated as $I = \rho_l v_l / \rho_u v_u$, where ρ and v refer to the soil density and wave velocity, respectively (the subscripts, l and u, refer to the lower rock and upper soil layer, respectively).

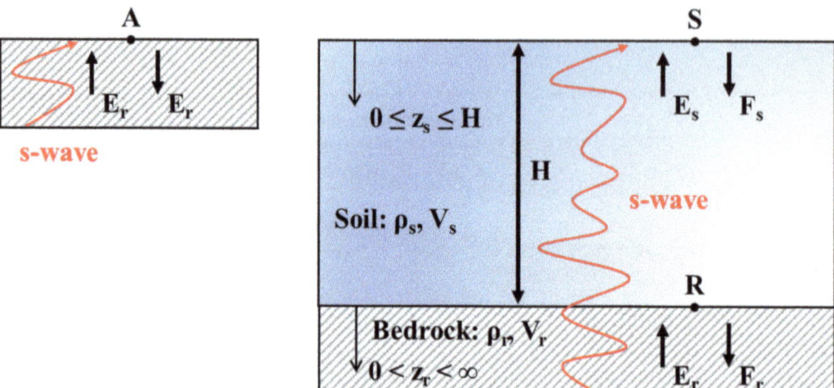

Fig. 1.9 Uniform, damped soil on elastic rock

$$\rho_s \frac{\partial^2 u_s}{\partial t^2} = G_s \frac{\partial^2 u_s}{\partial z_s^2} + \eta_s \frac{\partial^3 u_s}{\partial z_s^2 \partial t} \tag{1.8}$$

where ρ_s and G_s are the density and shear modulus of the soil material, respectively. The viscosity of the soil, η_s, can be defined by considering only one force–displacement cycle, as $\eta_s = (2G_s\xi_s)/\overline{\omega}$, where ξ_s is the damping ratio of the soil.[4]

By considering a possible solution as $u_s(z_s, t) = p_s(z_s)e^{i\overline{\omega}t}$, where $p_s(z_s)$ is the shape function that describes the $u_s(z_s, t)$ trend and provides its physical meaning, Eq. (1.8) becomes:

$$(G_s + i\overline{\omega}\eta_s)\frac{\partial^2 p_s}{\partial z_s^2} + \rho_s\overline{\omega}^2 p_s = 0. \tag{1.9}$$

Note that $\overline{\omega}$ is independent of both materials (soil and rock) and refers to the external seismic wave. By introducing the frequency-independent complex shear modulus, G_s^* $(= G_s + i\overline{\omega}\eta_s)$, and the complex wave number, k_s^* $(= \overline{\omega}/v_s^*$, where v_s^* is the complex shear wave velocity),[5] it is possible to obtain:

$$p_s(z_s) = E_s e^{ik_s^* z_s} + F_s e^{-ik_s^* z_s} \tag{1.10}$$

$$u_s(z_s, t) = \left(E_s e^{ik_s^* z_s} + F_s e^{-ik_s^* z_s}\right)e^{i\overline{\omega}t} \tag{1.11}$$

---

[4] The parameter, ξ_s, quantifies the material damping that reduces the response of the soil. This effect account for the non-linear dissipation of the soil due to the friction of its grains. Note that the model is linear elastic, thus it is assumed that for very low values of the damping ratio (i.e., $0 < \xi_s < 3.0\%$) the model maintains linear elastic. In fact, only one force–displacement cycle could be not sufficient to exceed the elastic limit.

[5] The complex wave number, k_s^*, is formed by a real, $k_{s,1}^*$, and imaginary, $k_{s,2}^*$, part, like the relation for the complex shear modulus, G_s^*.

where E_s and F_s are the amplitudes of waves traveling in the $-z_s$ (upward) and $+z_s$ (downward) directions, respectively, which represent the unknowns of the problem to be found (see Fig. 1.9).

Other hypothesis of this model considers that the wave amplitudes that pass through the rock-soil interface are partially reflected thus a part of energy is lost (e.g., $E_s \neq E_r$), however the general concept is that most energy is trapped in the soil material (i.e., $E_s \approx F_s$).

In a similar way it is also possible to obtain:

$$u_r(z_r, t) = \left(E_r e^{ik_r^* z_r} + F_r e^{-ik_r^* z_r}\right) e^{i\overline{\omega}t} \tag{1.12}$$

where, as already mentioned, the subscript "r" refers to the elastic rock, therefore all parameters in Eq. (1.12) have the same meaning as for soil, thus for brevity, they are not repeated here.

To respect the compatibility of the displacements (u_s, u_r) and the continuity of the stresses (τ_s, τ_r) at the free surface (point A and S) and at the soil-rock boundary (point R), some conditions are requested. In particular at S point, $\tau_s(z_s, t) = \tau_s(0, t) = G_s \gamma_s(0, t) = G_s u'(0, t) = 0$, where γ_s is the shear strain of the soil,[6] thus it is possible to obtain in Eq. (1.11) $E_s = F_s$, and:

$$u_s(z_s, t) = 2E_s \frac{\left(e^{ik_s^* z_s} + e^{-ik_s^* z_s}\right)}{2} e^{i\overline{\omega}t} = 2E_s \cos\left(k_s^* z_s\right) e^{i\overline{\omega}t}, \tag{1.13}$$

which represents the horizontal dynamic displacement of the soil (usually the most important parameter to be estimated) under a steady-state wave of the amplitude $2E_s \cos(k_s z_s)$, considering only the real part.

To solve Eq. (1.8) they are necessary to define the initial and boundary conditions, respectively:

$$u_s(z_s, 0) = 2E_s \cos(k_s z_s), \quad \frac{\partial u_s}{\partial t}(z_s, 0) = \overline{\omega}[2E_s \cos(k_s z_s)] \tag{1.14}$$

$$u_s(H, t) = 2E_s \cos(k_s H) e^{\overline{\omega}t}, \quad \frac{\partial u_s}{\partial z_z}(0, t) = 0. \tag{1.15}$$

Figure 1.10 shows some results in terms of shear stress and strain as example (developed by Wolfram Mathematica software). The analytical solution is described by the mentioned relations, the numerical solution can be retrieved in [13, 76], and the maximum values can be estimated by Deepsoil software. Here the agreements between all results appear good; the differences between the curves could be due to the estimation of E_s amplitude and adopted mechanical parameters for soil and rock.

Given that this model could simulate in a good way the behaviour of a real soil, it provides some practical relations to be used for a preliminary analysis, i.e., the

---

[6] Prime (′) denotes differentiation with respect to a geometrical coordinate, here it refers to the vertical coordinate z_s.

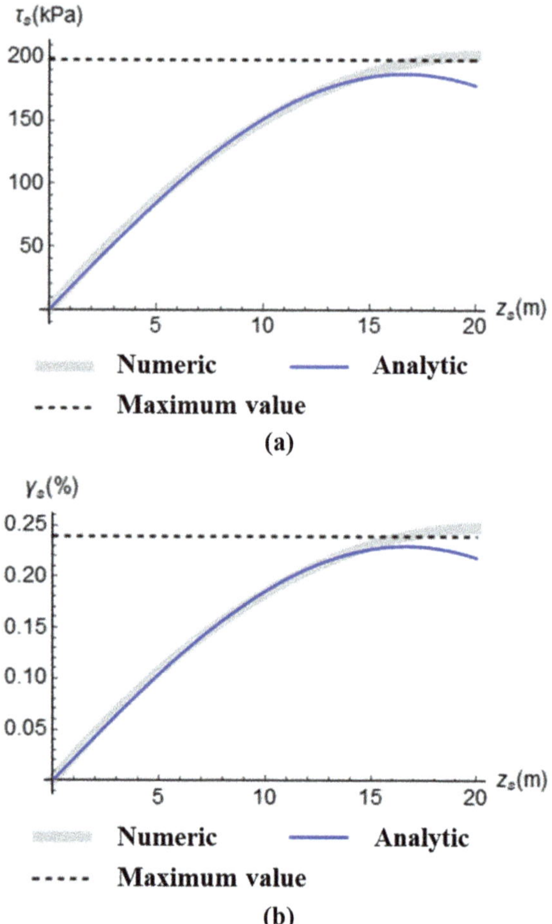

Fig. 1.10 Example of results for **a** shear stresses and **b** shear strains of a soil layer H

amplification peaks of the soil, A_1^*, regarding its fundamental frequency, f_1, [5][7]:

$$A_1^* \approx \frac{2}{\pi \xi_s}, \quad f_1 = \frac{v_s}{4H}. \tag{1.16}$$

However, it is important to highlight that the behaviour of a real soil can be altered by other phenomena. In fact, a real soil is formed by an heterogenous soil with a non-linear behaviour. The former is characterized by non-constant velocity, v_s, whereas the latter by high energy loss of the soil (i.e., by high ξ_s values).

In a very general way, the heterogeneity tends to amplify the soil response generating the amplifications. This is because in a heterogeneous soil the velocity,

---

[7] Also in this case, the asterisk indicates the complex phenomenon, thus the peak amplification A_1^* is composed by a real and imaginary part, for instance $A_1^* = A_{1,1} + iA_{1,2}$.

v_s, reduces approaching the surface thus its frequency, approximating to the soil frequency; in other words, the amplification peaks of the soil occur for low frequencies. On the other hand, a non-linear behaviour of the soil tends to dampen its response.

Finally, two classical schemes can be adopted to estimate the surface displacements for real soil by numerical analysis. The first one uses "continuous layers" where each layer follows the "uniform, damped soil on elastic rock" model, thus the mentioned relations can be used. Each layer could assume a maximum height of: $H_{max} \approx v_s/(7 \times \bar{f})$. In this model, the impedance ratio, I, assumes an important role since it quantifies the amplification between several layers; its values can be I $= 1.25$–10.0 [5], up to I $\rightarrow \infty$ (i.e., $(\rho_l v_l) \rightarrow \infty$) for rigid rock.

Another scheme uses the "lumped masses" where each layer is defined by a specific mass, stiffness and damping. This scheme is analogous to that used for the structures treated in Chap. 2.

1.3 Magnitude

The magnitude is the first parameter used to define the "size" of an earthquake. It is used in scientific filed but also in more common and popular dissemination. Given that, as discussed, there are several ways to register an earthquake due to different focal mechanisms, distances, waves, etc., several magnitudes exist. The more common magnitudes are described as follows [51]:

(i)   Local magnitude, M_L (or Richter magnitude), which depends on the registered peak amplitude in terms of displacements, A_L, with respect a fixed reference peak called "Wood-Anderson" amplitude, A_{L0}:

$$M_L = \log_{10}(A_L) - \log_{10}(A_{L0}), \tag{1.17}$$

which is usually reliable for low-medium magnitude, i.e., $M_L \approx 3.50$–5.50.

(ii) Surface wave magnitude, M_s (usually referring to Rayleigh wave), which is based on the maximum amplitude/duration ratio (A_s/T ratio) in terms of displacements and long periods (~ 20.0 s), and epicentral distance, Δ ($\equiv$ M2 in Fig. 1.2):

$$M_s = \log_{10}(A_s/T)_{max} + 1.66 \log_{10}(\Delta) + 3.30, \tag{1.18}$$

which is reliable for $M_s > 5.50$. The constants 1.66 and 3.30 have been calibrated experimentally when the relation was estimated.

(iii) Body wave magnitude, m_b (usually for p-wave), which is calculated by the amplitude/duration ratio (A_b/T ratio) for a short period (~ 1.0 s), and an empirical value in function of the epicentral distance and deep depth (> 50.0 km), $Q(\Delta, h)$:

$$m_b = \log_{10}(A_b/T) + Q(\Delta, h), \qquad (1.19)$$

which is more suitable for low-medium magnitude.

(iv) Moment magnitude, M_w (or M),[8] which is correlated to the seismic moment, M_0, depends on the rupture strength of the material along the fault, rupture area, and the average amount of slip. The magnitude M_w can be expressed empirically as:

$$M_w = \frac{2}{3} \log_{10} M_0 - 10.70. \qquad (1.20)$$

The seismic moment M_0 is expresses as a force times a length, thus it quantifies a mechanical energy.[9] For this reason, M_w is considered the best magnitude because, the definition of an earthquake size direct by its energy released from the source avoids "contaminations" of the soil, wave path, etc. Also, it is valid for both small and large magnitudes (3.50–8.0).

Several variations of these magnitudes exist. For instance, to estimate the local magnitude it is possible to use the regional magnitude, m_R (used for short p-waves) [8], and the Japanese meteorological agency magnitude, M_{JMA} (used for long waves). An extension of the M_s can be the Lg-wave magnitude, M_{Lg}, which is calculated like M_s but considering the Lg-waves [6]. A variation of the magnitude M_w is the magnitude M_{ww}, which accounts for the "w-phase" that observes at large distances ($> 500.0\,$km) for waves with very long period (200.0–$1000.0\,$s); it is useful to estimate strong earthquakes.

Other magnitudes are the coda magnitude, M_c (for backscattered waves), duration magnitude, M_D (for small earthquakes), etc.

All these large varieties indicate that the estimation of the magnitude is subjected to uncertainties usually quantified around $\pm\ 0.20$. In general, earthquakes with a magnitude less than 3.0 are recorded by seismographs, but are usually not felt by people and structures. Magnitudes between 3.0 and 4.9 are often felt but rarely cause damage. Earthquakes with magnitudes between 5.0 and 6.9 are considered moderate to strong and can produce damage on structures causing deaths and injuries. Magnitudes greater than 7.0 are classified as major to great earthquakes, with significant potential for widespread destruction.

---

[8] Given that the magnitude M_w is considered the best magnitude, thus currently the most used since the 1980s, to define the size of an earthquake, it is possible to find in literature only the letter "M" instead of M_w, which is more immediate and easier to understand for everyone. Previously, the M_L was used more (since the 1940s). However, this is a general indication.

[9] The used model to estimate M_0 is correlated to the Brune's model (described in Chap. 6), which is based on the Fourier spectrum in terms of displacements of the fault. However, due to the enormous distance between the fault and the site, it is more difficult to compute it in an exact way.

Chapter 2
Elastic Response Spectrum

Abstract The elastic response spectrum method is provided in all national and international codes, making it the primary approach for designing buildings, civil works, and industrial structures worldwide. Although relatively simple to apply, it requires careful calculation of certain parameters. This chapter details the parameters and procedures of spectral response (both general and code-specific) which are often not directly explained in the codes themselves.

This method is provided by all national and international codes. For this reason, it represents the main method to be used for buildings, civil and industrial structures in the world. It is a fairly simple method to use, however there are some parameters that must be calculated with a particular attention. Here some aspects and concepts are explained that are usually not explained in codes.

2.1 Spectral Response

A structure to be designed for earthquake actions must be modelled in a satisfactory and adequate way. The general idea of this method is to schematize a structure as a simple oscillator with a single degree of freedom (SDOF) system. A simple oscillator is an ideal structure with at the top there is a unique mass, which is fixed with the foundation by a vertical element with a certain stiffness and damping. An earthquake, by moving the soil, transmits its vibrations through this vertical element reaching the top mass where a fictitious horizontal force is applied to quantify the structural displacements.

The concept of a horizontal force that produce a displacement leads directly to an energy concept, which is treated in several parts of this book. This practical use is also consistent to the fact that the inertia force is proportional to the mass.

Figure 2.1 shows a SDOF system with unique mass and two columns; even in this case it can be called "simple oscillator" since the mechanical concept maintains.

E. Zacchei and R. M.L.R.F. Brasil, *Calculation of Seismic Actions for Structural Analysis*, https://doi.org/10.1007/978-3-032-08884-0_2

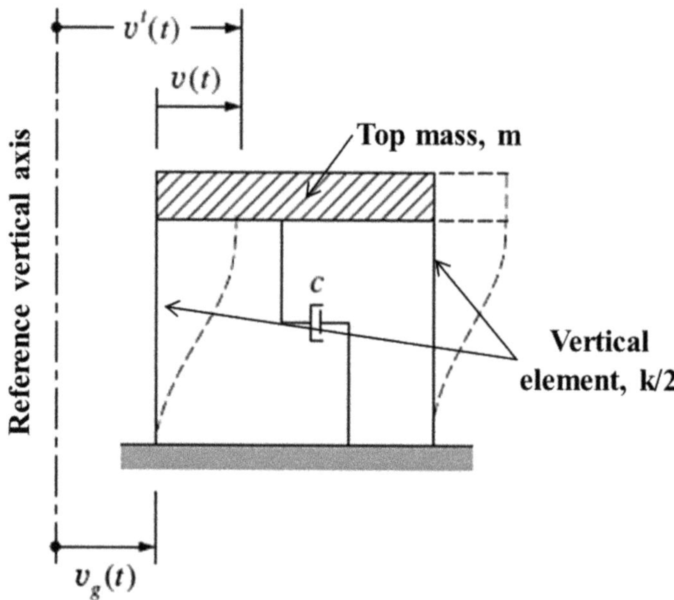

Fig. 2.1 Single degree of freedom (SDOF) dynamic system (adapted from [15])

This ideal model of the simple oscillator has been conceived since it represents the behaviour of several real structures. For instance, an elevated concrete storage tank (see Fig. 2.2a) has a great mass at the top of the structure (usually containing liquid), which is supported by vertical piles with a relatively small mass designed for compressive and flexural stresses. The beams and piles of a reinforced concrete bridge (see Fig. 2.2b) can be also represented by several simple oscillators where the half of the right and left beam is the top mass, and the piles are the vertical elements of the oscillator.

Other structures can be modelled in this way, as for instance a pile supported wharf [18, 27], a building [16], etc. However, it is important to highlight that this ideal model, in many cases, provides very approximated results thus it should be used only for preliminary analyses.

The equation of motion for this simple system is most easily formulated by directly expressing the equilibrium of all forces acting on the mass using d'Alembert's principle (based on the Newton's second law of motion and originally introduced by Galilei [80]). The concept of this principle is that a mass develops an inertial force proportional to its acceleration and opposing it.

The forces acting in the direction of the displacement degree of freedom, in the time domain, t, are external inertial load due to the horizontal ground acceleration, $\ddot{v}_g(t)$, and the three resisting internal forces resulting from the motion, i.e., the inertial force, $m\ddot{v}(t)$, the damping force, $c\dot{v}(t)$, and the elastic spring force, $kv(t)$. Thus, the equation of the motion can be expressed by:

(a)

(b)

Fig. 2.2 Real photos of **a** an elevated concrete storage tank and, **b** a reinforced concrete bridge (by iStock site)

$$m\ddot{v}(t) + c\dot{v}(t) + kv(t) = -m\ddot{v}_g(t) \tag{2.1}$$

where m is the entire mass of the system, c is a damper that represents the energy loss mechanism, and k is the weightless spring of stiffness that provides the elastic resistance to displacement. The single horizontal displacement coordinate is defined by v(t), where $\dot{v}(t)$ and $\ddot{v}(t)$ are the first and second time derivate of v(t), i.e., the horizontal velocity and the acceleration of the system, respectively.

By dividing by, m, and introducing some relation, as for instance,

$$\omega^2 \equiv \frac{k}{m}, \quad c \equiv 2\xi\omega m. \tag{2.2}$$

Equation (2.1) can be expressed in a more compact way:

$$\ddot{v}(t) + 2\xi\omega\dot{v}(t) + \omega^2 v(t) = -\ddot{v}_g(t). \tag{2.3}$$

Note that Eq. (2.2) indicates identical (i.e., the symbol "$\equiv$") relations to express Eq. (2.3) in a more convenient way in order to consider only the positive part and, in any case, the imaginary part has been neglected considering only the real one. The parameter ω represents the undamped circular frequency of the system, whereas ξ is the damping ratio of the system. This ratio is calculated between the damper, c, and its critical value.

The solution of Eq. (2.3) is expressed by the Duhamel integral[1] in the t-domain:

$$v(t) = \frac{1}{m\omega_D} \int_0^t -m\ddot{v}_g(\tau) \sin \omega_D(t - \tau)e^{[-\xi\omega(t-\tau)]}d\tau \tag{2.4}$$

where ω_D is the free vibration frequency of the damped system, and τ is a defined short time. The parameter, ω_D, is the damped circular frequency of the system, which is directly correlated with ω and ξ, by a circular function. In particular, when the system is totally damped thus $\xi = 1.0$ (i.e., 100.0%), $\omega_D = 0$, whereas when the system is not damped thus $\xi = 0$, $\omega_D = \omega$.

These values represent two ideal limits since a real structure cannot dissipate energy completely, given that a part of internal response is also caused by inertial and spring force. In this sense, all three terms at the left-hand side of Eq. (2.3) play a role for the structural response. For a very stiff structure it is possible to neglect the first two term and to define the response by $\omega^2 v(t) \approx -\ddot{v}_g(t)$.

It is not even possible that a real structure does not dissipate no energy, since the dissipation is intrinsic to the characteristics of the materials that a structure is made of. In this case, the approximated response could be defined by $\ddot{v}(t) \approx -\ddot{v}_g(t)$, which corresponds to a very flexible structure.

In general, for civil and industrial structures, the damping ratio assumes a small values, for example between $\xi = 2.0\%$ for steel structures up to $\xi = 10.0\%$ for elasto-plastic reinforced concrete structures when the hysteretic (dependence on its cyclical history) behaviour of the material is considered. Therefore, it is possible to assume $\omega_D \approx \omega$. This approximation is also valid for $\xi = 20.0\%$, however this value concerns more mechanical structures and applications.

Under this approximation, $\omega_D \approx \omega$, and neglecting the negative sign in Eq. (2.3) since it has no real significance regarding the earthquake excitation, it is also possible

---

[1] The Duhamel integral has been obtained by considering the response of an undamped SDOF structure under short duration impulsive loads. From this, the impulsive loads have been extended to model a general dynamic loading by considering an arbitrary general loading, p(τ), and a trigonometric function, sin[$\omega(t - \tau)$].

to assume $v^t(t) = v(t) + v_g(t)$, where $v^t(t)$ is the total displacement of the structure-earthquake system by an external reference vertical axis (see Fig. 2.1).

It is usually useful to estimate only the pseudo-velocity spectral response, $S_{pv}(\xi, \omega)$, and the pseudo-acceleration spectral response, $S_{pa}(\xi, \omega)$, defined by, respectively:

$$S_{pv}(\xi, \omega) \equiv \left[\int_0^t \ddot{v}_g(\tau) \sin \omega(t - \tau) e^{[-\xi\omega(t-\tau)]} d\tau \right]_{max} \tag{2.5}$$

$$S_{pa}(\xi, \omega) = \omega S_{pv}(\xi, \omega), \tag{2.6}$$

which is particular important since, $S_{pa}(\xi, \omega)m$, provides the maximum inertial force developed in the simple oscillator under an earthquake, $F_{s,max}$ (i.e., $S_{pa}(\xi, \omega)m = F_{s,max}$).[2]

Another relation using $S_{pv}(\xi, \omega)$ is: $S_{pv}(\xi, \omega) = \omega S_d(\xi, \omega)$, where $S_d(\xi, \omega)$ is called "spectral relative displacement". Note that the term pseudo-velocity and pseudo-acceleration is justified to the fact that a dynamic behaviour of the oscillator is simulated by a static analysis where the frequency, ω, would replace the time, t. In fact, it is known that the unique difference between a static and dynamic analysis is the direct dependency of the time, t.

Structural engineers normally need to evaluate the structural responses in terms of maximum values, however, often it is useful to know the characteristic features of such responses to generate the Fourier spectrum for the ground motion. As discussed later, the Fourier spectrum is direct correlated to the energy involved in the system, which is the best quantity (together to the displacements) to well evaluate the total responses (for more details regarding the Fourier spectrum see Appendix A in Chap. 8).

2.2 Elastic Response Spectrum for Code

The earthquake motion at a given point on the surface is simulated and modelled by an elastic ground acceleration response spectrum, more simply called "elastic response spectrum" as already explained.

In codes and standards, there are other aspects to be considered to obtain the more adequate elastic response spectrum. For instance, the shape of the spectrum, which is assumed the same for two levels of seismic action, i.e., for (i) no-collapse and (ii) for damage requirements.

(i) The no-collapse requirement belongs to the ultimate limit state (ULS); thus, it implies that the structure shall be designed and constructed to withstand the

---

[2] It is possible to demonstrate mathematically that considering $\xi = 0$, Eq. (2.6) provides a satisfactory response, and the following approximation maintains valid: $\dot{v}(t) \approx S_{pv}$. Due to this approximation, from Duhamel integral it is obtained the practical pseudo-acceleration spectral response, $S_{pa}(\xi, \omega)$.

design seismic actions without local or global collapse, retaining its structural integrity and a residual load bearing capacity after the seismic events.

This design seismic action is expressed in terms of:

(a) Reference return period, T_{NCR} (for standard civil and industrial structures, $T_{NCR} = 475.0$ years). This value is estimated by the following general relation: $T_R = -T_L/\ln(1 - P_R)$, where T_R is the requirement return period, P_R is the requirement probability of exceedance in T_L years of a specific level.[3] For a non-collapse requirement (NCR) for standard structures under a liner elastic behaviour, by assuming recommended values of $P_R \to P_{NCR} = 10.0\%$ and $T_{LR} = 50.0$ year, the value of $T_R \to T_{NCR} = 475.0$ years is obtained.

(b) The importance factor, γ_I, considers reliability differentiation, which relates to the consequences of a structural failure, and to its specific use (it is also common to adopt the terminology of "important class" for this purpose). This factor times the reference peak ground acceleration (PGA), here a_{gR}, produces the design ground acceleration, a_g, described in Eqs. (2.7)–(2.10) (i.e., $a_g = \gamma_I a_{gR}$). This factor can be estimated as $\gamma_I = (T_{LR}/T_L)^{-1/k}$, where T_{LR} is the reference period that is correlate to the specific use of the structure ($T_{LR} < T_L$ for large structures, structures used for large aggregations of people, etc., in order to provides $\gamma_I > 1.0$), and the exponent k depending on the seismicity (usually $k = 3.0$).[4] For standard structures it is possible to assume that $\gamma_I \approx 1.0$. It is also possible adjust T_L or γ_I to avoid considering this phenomenon twice thus to overestimate the design forces.

(ii) The damage requirement belongs to the serviceability limit state (SLS); thus, the structure shall be designed and constructed to withstand a seismic action having a larger probability of occurrence than the design seismic action, without the occurrence of damage and the associated limitations of use, the costs of which would be disproportionately high in comparison with the costs of the structure itself. Thus, for a damage limitation requirement (DLR) for temporary structures under a liner elastic behaviour, by assuming recommended values of $P_R \to P_{DLR} = 10.0\%$ and $T_{LR} = 10.0$ year, the value of $T_R \to T_{DLR} = 95.0$ years can be used.

It is visible how it is possible to work in terms of probability, P_R, or period, T_R, values (in general, working in terms of return period is easier for users to understand). However, to use only the T_R value as "reference" parameter to carry out a seismic analysis could be inadequate. This is because to correlate the importance of the structure with the return period of the event is conceptually wrong, since the former regards the use and the typology of the structures, which does not intrinsically specify

---

[3] This value represents the design life of the structure, which can be assumed: $T_L = 10.0$ years (for temporary structures), $T_L = 50.0$ years (for standard structures), $T_L = 100.0$ years (for large structures) [17].

[4] This parameter measures the level of seismicity, where a great value of k indicates a high seismicity of the zone. The k-parameter is similar conceptually to the a-value treated in Chap. 5.

its mechanical characteristics (e.g., a school building can be low and wide or high and slender), whereas the latter regards an external seismic event that can occur in any points and with any magnitude in the Earth independently of the typology of the structure on which it acts.

To correlate the importance of the structure with the design ground acceleration could be also misleading since a very high important structure (e.g., hospitals, fire stations, etc.) could have small dimensions thus an amplification in favour of safety could greatly overestimate the seismic design forces.

Some current approaches correlate the importance factor to the risk analysis, which is a holistic analysis divided into three main categories: hazard (i.e., the adverse event causing the loss), exposure (i.e., the property, people, environment that are threatened by the event), and vulnerability (i.e., how fragile is a structure to an adverse event). These approaches appear completer and more adequate (for more details see [20, 21]).

Another aspect regarding the elastic response spectrum is the components of the seismic waves thus the seismic actions (treated in Chap. 1). The horizontal seismic action is described by two orthogonal components (in the plane of the ground) assumed as being independent each other and that they can be represented by the same response spectrum. For a special model in three components (two horizontals and one vertical) of the seismic action, the vertical accelerations range between 0.45 and $0.90a_g$ (for example in the Venezuelan code a unique mean value of $0.70a_g$ is assumed [22], however this code is not the unique existing example).

The three components of a seismic action generate different accelerations; however, both horizontal components are assumed the same, since it is not possible a-priori to place the structure according to the direction of an expected earthquake; also, the arrival of an earthquake in one direction rather than another one can generate added torsion moments and rotational accelerations (or "rocking accelerations" [39]), not considered during the design (especially for irregular structures in height and plan [25], or for unanchored storage tanks on ground [72, 81]). The vertical component can be neglected due to its very low value; however, it is necessary to verify the indications of the used code.

The elastic response spectrum in terms of accelerations, $S_e(T)$ (analogue to the previous parameter $S_{pa}(\xi, \omega)$ in Eq. (2.6), whit $\omega = 2\pi/T$), for both horizontal components of the seismic action is defined by:

$$S_e(T) = a_g S \left[1 + \frac{T}{T_B}(\eta 2.5 - 1) \right], \quad 0 \le T \le T_B \tag{2.7}$$

$$S_e(T) = a_g S \eta 2.5, \quad T_B \le T \le T_C \tag{2.8}$$

$$S_e(T) = a_g S \eta 2.5 \left[\frac{T_C}{T} \right], \quad T_C \le T \le T_D \tag{2.9}$$

$$S_e(T) = a_g S \eta 2.5 \left[\frac{T_C T_D}{T^2} \right], \quad T_D \le T \le 4\,\text{s} \tag{2.10}$$

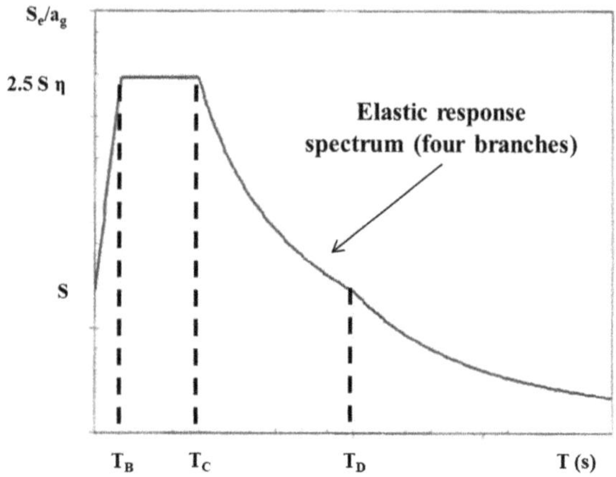

Fig. 2.3 Shape of the elastic response spectrum (adapted from [14])

where T is the vibration period of a linear SDOF system, T_B and T_C is the lower and upper, respectively, limit of the period of the constant spectral acceleration branch, T_D is the value defining the beginning of the constant displacement response range of the spectrum,[5] S is the soil factor, and η is the damping correction factor defined in function of the damping ratio, ξ, as $\eta = \sqrt{10/(5+\xi)} \geq 0.55$.

As mentioned, for standard reinforced concrete buildings with $\xi = 5.0\%$,[6] $\eta = 1.0$; whereas for special cases as for example non-linear reinforced concrete structures with $\xi = 10.0\%$, $\eta = 0.82$ thus a further damping is considered. To estimate better the value of ξ ratio it is necessary to carry out experimental laboratory tests since the values provided by codes are approximative; however, the additional resistance due to the elasto-plastic behaviour of the reinforced concrete structures, prestressed concrete structures, and steel structures compensates for this approximation, therefore its underestimation should not cause damage or collapse to the structure.

Figure 2.3 shows the classical shape of the elastic response spectrum plotting by using Eqs. (2.7)–(2.10). The shape of the elastic response spectrum came from to the numerical procedure explained in Chap. 3.

Note that the elastic response spectrum is composed by four branches where the points T_B, T_C, and T_D define the limits of each branch. In the seismic codes in the World, it is also possible to find elastic response spectrum by two and three branches, as for instance in the Peruvian [23] and Indian code [24] where the first and fourth branch is neglected, respectively. These differences are due to the level of the approximation that the specialists of each country decide to assume.

---

[5] The "beginning of the constant displacement response" is a point of a spectrum obtained by curves on the four-way log plot where the maximum spectral displacement, velocity and acceleration is plotted. This four-way regards the three vertical axes and one horizontal axis. This method is known as Newmark-Hall method (for more details see [15]).

[6] The design elastic spectrum is usually plotted with a damping of $\xi = 5.0\%$, since the seismic codes are prepared for reinforced concrete buildings. Despite the presence of a possible damping due to the material composition, the structure maintains a linear elastic behaviour (actually, $\xi = 5.0\%$ is a very low value).

The first branch includes very low values of the structural period (T < 0.20 s), which are not common for instance for high buildings. A structure infinitely stiff moves with the soil thus the relative displacement of the structure is null. The fourth branch includes high values of T (T > 1.20 s), which can be representative of flexible structures as for example bridges, pile supported wharf, etc. In this case an adequate calculation of T is necessary, since the flexibility of the structure plays an important role with respect the soil-structure interaction altering the relative displacement of the structure. However, neglecting the first and the fourth branch, the values of the spectral accelerations are more conservative providing higher values; in this sense, each standardization spectrum can be substantially considered in favour of safety.

The spectral accelerations, $S_e(T)$, for buildings usually do not exceed the values of T > 4.0 s, for this Eq. (2.10) has this upper limit.

The design ground acceleration, a_g, is considered for a rock ground where there are not amplification phenomena due to site effects (i.e., variation of the composition of the stratigraphic profile. In Eurocode they are classified as soil A, B, C, D [14]), topographic effects (i.e., presence of topographic irregularities), and effects of basin geometry (i.e., presence of alluvial valleys. In Eurocode it is classified as soil E [14]). In function of the ground type where the structure is placed, the value of the soil factor, S, changes. The type of soil intrinsically also includes the heterogeneity effects showing by the variation of the average shear wave velocity (discussed in Chap. 1) along the depth of a specific stratigraphic profile. In general, all these effects tend to amplify the accelerations, however they could also reduce them.

The a_g value can be modified for specific cases under adequate studies and justifications (usually this analysis is requested for large important structures). It is possible to take different values of the acceleration, a_g, when the earthquakes that affects a site are generated by widely differing seismic sources (explained in Chap. 5).

Conceptually, the elastic response spectrum can vary in three ways:

(i) Curves subjected to the soil effects (see Fig. 2.4a): The curves move from left side to right side thus from a structural response in a rock soil (called "soil A") to a response in a soft soil (soil D), respectively [14]. The variation of the composition of the stratigraphic profile more alters the maximum spectral velocity (points in $T_C \leq T \leq T_D$) and maximum spectral displacement (points in $T \geq T_D$). The spectral accelerations suffer minor variations. In general, a flexible structure in flexible soil generates great soil-structure interaction. However, if a stiff structure is supported on a relatively soft soil, considerable energy will be transferred from the structure to the soil, and the base motions will differ drastically from those experienced by the soil under free-field conditions. A rock soil does not interact with the dynamic response of the structure. The response of the structure, idealized as a simple oscillator, is represented only by the points for T > 0, since at T = 0 the PGA only regards the information

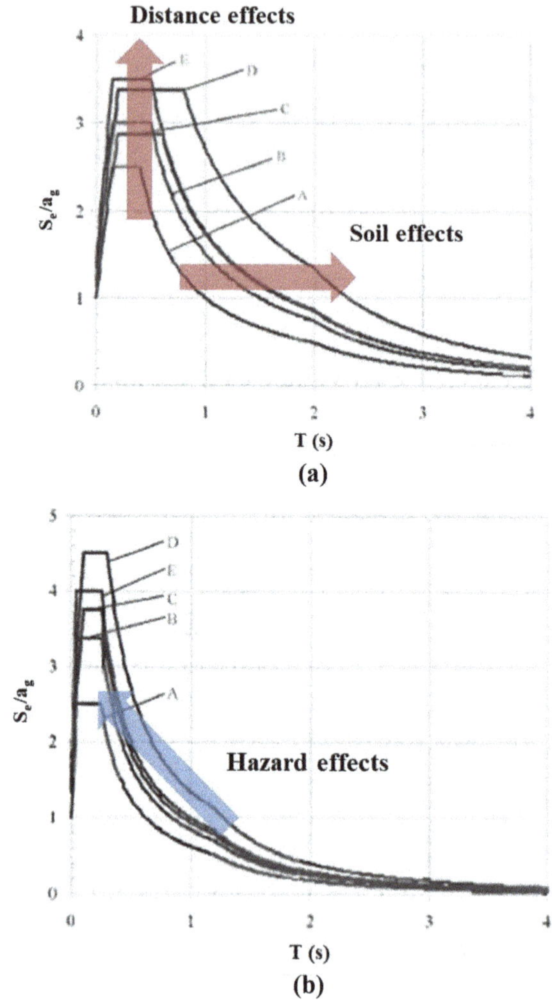

Fig. 2.4 Shape of the elastic response spectrum (5.0% damping): **a** Type 1 (for $M_s > 5.5$), and **b** Type 2 (for $M_s < 5.5$) (adapted from [14])

of the soil.[7] Note that the presence of the alluvial layers (soil E) generates a different amplification phenomenon more similar to the following (ii) effect.

(ii) Curves subjected to the distance effects (from the seismic source to the structure, Fig. 2.4a): In general, for the near-field earthquake, the spectral acceleration and velocity increase, due to long-term frequency (i.e., short period) of the seismic waves. This aspect can produce resonance effects for structure with high frequency (i.e., fundamental frequency of the structure $\approx$ frequency

---

[7] Rigorously, in accordance with the response spectrum curves on the four-way log plot, the first branch between $0 \leq T \leq T_B$, is plotted to purely joint the PGA with the maximum spectral acceleration. The structural response of a dynamic action (earthquake) on a static scheme (simple oscillator) can be estimated approximately as 2.0 times of the static response [19]. In this code the value of 2.50 represents an analogue concept.

of the earthquake[8]). On the other hand, the great earthquakes are characterized by long periods, thus affecting more flexible structures. Also, the superficial waves (see Chap. 1), which affect directly the structure, have usually large periods.

(iii) Curves subjected to the seismic hazard effects (Fig. 2.4b): If the earthquakes that contribute most to the seismic hazard defined for the site for the purpose of probabilistic seismic hazard analysis[9] have a surface-wave magnitude, M_s, not greater than 5.5, it is recommended that the Type 2 spectrum is adopted. Otherwise, for $M_s > 5.5$ it is recommended to adopt the Type 1 spectrum. In general, earthquakes with $M_s < 5.5$ are characterized by long frequencies affecting stiff structures. Here it is possible to see that the great earthquakes, which produce non-linear deformations, generate high dissipations thus low amplifications due to the soil, S; in fact for the Type 1 the vertical axis is lower than that for Type 2. An order of magnitude of a "alarming" value of PGA could be $PGA = 0.10g$, i.e., an increasing of the element masses of a structure of 10.0%.

The capacity of a structure to resist seismic actions in the non-linear range generally permits their design for resistance to seismic forces smaller than those corresponding to a linear elastic response. The capacity of the structure to dissipate energy, mainly through the ductile behaviour of its structural and non-structural elements and members, is considered by the behaviour factor, q, which reduces the elastic response spectrum providing the "design spectrum". This q factor substantially substitutes the η factor (i.e., $\eta \equiv 1/q$). This factor mainly depends on the type of the structural system, its regularity in elevation, flexural resistance, capacity to generate plastic hinges, and its prevailing failure mode (e.g., brittle or ductile).

For the horizontal design spectrum, the factor q assumes value $q \geq 1.50$, whereas for the vertical design spectrum is $q < 1.50$ since buildings have a geometrical form that allows little dissipation of energy in a vertical direction. In general, it is not simple to define a correct q value, for this in many cases it is necessary to adopt specific non-linear models, experimental laboratory tests, advanced numerical analyses, etc.

A plastic hinge represents a region of a structural element where high stresses are expected. The concept is that a structure must maintain a linear elastic behaviour during the seismic actions, however if some elements accumulate permanent deformations thus assuming a plastic behaviour it is better that this accumulation occurs only where the plastic hinges have been placed during modelling to dissipate the energy [26]. Usually, they are placed where the flexural bending generates maximum

---

[8] An earthquake is characterized by infinity frequencies; thus, here the "frequency of the earthquake" refers to the frequency that provides the maximum amplitude in the Fourier spectrum for a certain earthquake. It also exits a phenomenon of "double resonance" between earthquake-soil and soil-structure.

[9] This probabilistic seismic hazard analysis is treated in Chap. 5. Note that the elastic response spectrum by codes is deterministic, since Eqs. (2.7)–(2.10) do not directly consider probability distributions, rate of exceedances, stochastic processes, etc. Thus, correlating it to a probabilistic approach could be misleading.

values, and they consist in adding a great number of longitudinal steel bars and steel stirrups in the concrete.

Chapter 3
Simplified and Direct Methods

Abstract This chapter presents simplified and direct methods, which are easy to apply but overlook several important aspects. The simplified method, though now considered outdated, is based on universal mechanical principles that remain valid, such as the application of inertial forces on a structure. Direct methods, by contrast, derive directly from the time-history records of real earthquakes, providing a more accurate representation of seismic response.

The methods described in this chapter (simplified method and directs methods) are very easy to be applied and consequently they neglect several important aspects. The simplified method nowadays is considered incorrect; however, it follows mechanical universal principles (such as the principle of inertia) that are still valid. The direct methods are called in this from since they derivate directly from the registration of a real earthquake.

3.1 Seismic Coefficient (Simplified) Method

The simplified method here called "seismic coefficient method" is no longer used due to excessive approximations mainly in terms of the seismology. However, it could be interesting to mention it to understand the mechanical concept in applying a static inertial force in a structural element, which is always valid. In fact, in some codes are still present since they can provide an efficient preliminary seismic calculation as for instance in Spanish code [32].

For small structures, a static analysis for earthquake design could be satisfactory. This analysis approximates the dynamic loads by a set of externally applied static forces that are applied laterally to the structure.

This method has been developed since the 1970s, when the seismic hazard maps were not yet well developed, and when there was the passage from the analogic system to the digital system. To explain this method, it is useful to do directly some examples.

© The Author(s), under exclusive license to Springer Nature Switzerland AG 2025

E. Zacchei and R. M.L.R.F. Brasil, *Calculation of Seismic Actions for Structural Analysis*, https://doi.org/10.1007/978-3-032-08884-0_3

In the old Italian code of the year 1975 [29], the seismic effects were evaluated by means of static analysis of the structure subjected to a system of horizontal forces, F_h, parallel to the direction expected for the earthquake estimated by:

$$F_h = CRW \tag{3.1}$$

where C is the seismic intensity coefficient that accounts for the "seismicity degree", R is the response coefficient estimated in function of the fundamental structural period, and W is total weight of the structural masses, which provides the physical meaning to Eq. (3.1). For a system of vertical forces, the calculation is analogue.

Note that Eq. (3.1) is valid not only for building but also for concrete dams, where the coefficient C assumes higher values. This application should demonstrate the enormous approximation in estimating a seismic force for very important and strategic structures.

It is possible to see that to calculate the seismic actions, two coefficients, C and R, have been adopted (for this reason this method is called seismic "coefficient" method), which substantially do not consider the seismicity of the site and other effects (e.g., source-site distance, soil characterization, hazard analysis, etc.). Also, this method does not consider in adequate way other aspects (e.g., foundation characterization, presence of the vertical stiff elements, etc.); these aspects are accounted for by coefficients showing further the simplicity of the method.

In American ASCE 7-10 code as explained in [75], the simplified method is based upon finding a seismic response coefficient, C_s, determined from the soil properties, the ground accelerations, and the vibrational response of the structure. For most structures, this coefficient is then multiplied by the mentioned total weight, W, to obtain the "base shear" in the structure. The value of C_s is actually determined from:

$$C_s = \frac{S_{DS}}{R_1/I_e} \tag{3.2}$$

where S_{DS} is the spectral response acceleration (explained in Chap. 2) for short periods of vibration, R_1 is a response modification factor that depends upon the ductility of the structure (steel frame members which are highly ductile can have a value as high as 8.0, whereas reinforced concrete frames can have a value as low as 3.0), and I_e is the importance factor that depends upon the use of the building (for example, $I_e = 1.0$ for agriculture and storage facilities, and $I_e = 1.50$ for hospitals and other essential facilities).

With each new publication of the code, values of these coefficients are updated as more "accurate" data about earthquake response become available.

Finally, in the old Portugues code in the year 1983 [28], it was provided the following Eq. (3.3) to estimate the characteristic forces, F_{ki}, to be applied at a building floor, i, up to a certain floor number, n:

$$F_{ki} = \beta h_i G_i \frac{\sum_{i=1}^{n} G_i}{\sum_{i=1}^{n} h_i G_i} \tag{3.3}$$

where h_i is height of the considered building floor, i, up to the total floor number, n, and G_i is the permanent and quasi-permanent loads of the floor. Here the seismic action is accounted for the seismic coefficient, β, which is estimated in function of the soil characteristics, structural frequency, ductility of the material, and seismic zone. The summation considers a unique global seismic force in the building distributed for each floor.

Like Eqs. (3.1)–(3.2), also in this last example the seismic action that depends on several complex phenomena is considered only by a simple coefficient.

3.2 Direct Methods

3.2.1 Time-History Representation

The seismic action may also be represented in terms of ground acceleration time-histories, and their relative velocities and displacements. In this sense, this method can be considered as a direct method where only a specific seismic database is necessary to retrieve the time-histories.

Figure 3.1 shows, as example, the t-histories for accelerations, velocities and displacements of the Turkey earthquake with moment magnitude M_w 7.6 and depth 17.0 km (event date 17/08/1999, latitude 40.75°, longitude 29.86°). The selected station, which has registered this event, "TK 8101"[1] is placed at a distance from the epicentre of 97.50 km [11].

The signals are recorded for the two orthogonal directions in the ground plane where the structure is placed, i.e., north (N), east (E), and for the vertical (Z) direction along the height of the structure. For the buildings, civil and industrial structures, both horizontal components are more dangerous in fact the maximum values of the accelerations are 367.79 cm/s^2 and 311.4 cm/s^2, regarding the E and N orientation, respectively. The vertical acceleration provides a maximum value of 191.39 cm/s^2. These maximum values of the accelerations are called peak ground accelerations (PGAs), already mentioned in Chap. 2. The maximum value for the velocity and displacement can be called peak ground velocity (PGV) and peak ground displacement (PGD),[2] respectively. The maximum PGV ad PGD values are 59.42 cm/s and 37.98 cm, respectively.

---

[1] Each seismic event has a unique hypocentre and epicentre, but it is registered by various stations (called "seismic instrument" in Fig. 3.2) placed in different points in the countries, which provide different displacement, velocity and acceleration values due to the soil characteristics where the station is placed. For this, it can be useful to select a station that provides maximum values or a station nearest to the studied structure to know the possible effects of the soil usually difficult to be estimated a priori.

[2] To differentiate the vertical components from the horizontal ones, it is possible to find in literature the acronyms HPGA and VPGA to identify the horizontal and vertical PGA, respectively. For PGV and PGD it is possible to use the analogue acronyms.

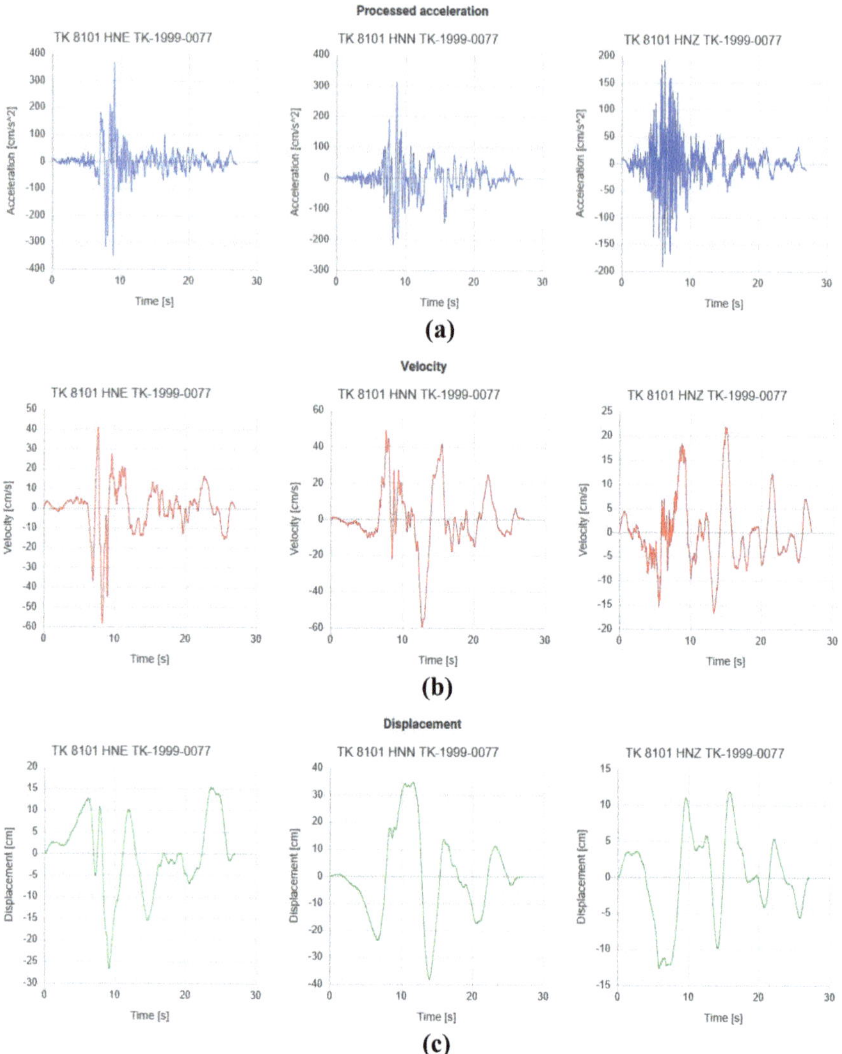

Fig. 3.1 Time-history representation in terms of **a** accelerograms, **b** velocities, **c** displacements of the Turkey 1999-year earthquake (Station code: TK 8101; instrument code: HNE, HNN or HNZ; event name: TK-1999-0077. Retrieved from the ESM database [11])

The displacement time-history is obtained by double integration in the time of the acceleration time-history. This mathematically operation can produce two mains physically incongruences:

(i)	The first incongruence regards the production of high values of the displacement at the end of the earthquake (for $t = 0$). The response must be an elastic response thus for $t = 0$ the amplitude of the displacements must be null.

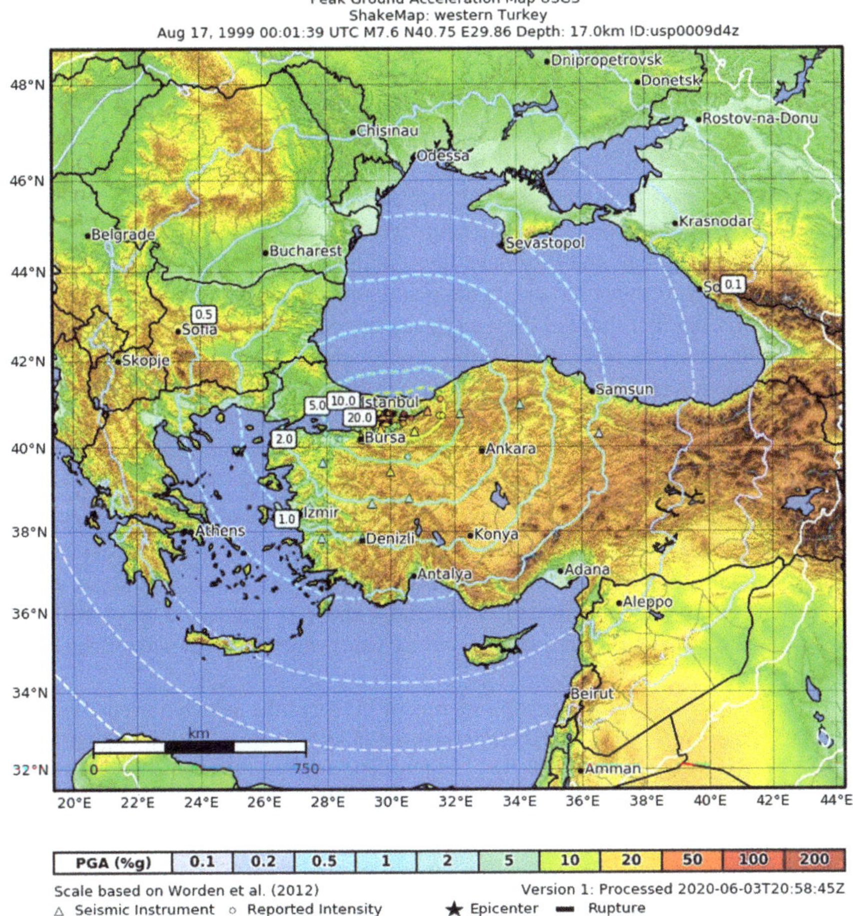

Fig. 3.2 Shake map in terms of PGAs of the Turkey 1999-year earthquake (adapted from [9])

(ii) The second incongruence regards the presence in the signal of high or small frequencies due to external noise, noise introduced by the digitization, unrolling and enlargement of the film, trigger effect, etc., in fact a seismogram records all type of vibrations and not just those of an earthquake.

Therefore, both incongruences affect the signal in the time-domain altering the time-histories and in the frequency-domain altering the Fourier spectrum. By using an adequate software, it is possible to adjust both incongruences through a tool usually called "baseline correction", which eliminates the mean value of the accelerations (null for a real stationary and ergonomic process as described in Chap. 4) thus the excess displacement, and through a tool "Butterworth filter"[3] between, for example, 0.20–25.0 Hz. This interval, related to the cut-off frequency of the high-pass filter, H_p, or of the low-pass filter, L_p, respectively, can be found "playing" with the adopted software up to obtain a processed acceleration consistent to the explained physical meanings.

In Fig. 3.1, the word "processed" in the acceleration signals, refer the mentioned adjustments, to pass from unprocessed accelerations to processed accelerations.

The acceleration trends are very different with respect the displacement ones since they are more irregular, and they have more noticeable vibrations. Mathematically, this difference is correlated to the mentioned double integration, whereas physically it depends on the fact that the displacements are cumulative, and the equations of compatibility must be respected. During the accelerations, at a certain time a part of a soil layer can move in a direction reaching a maximum value, in this sense, the displacements seem "slower" with respect the acceleration.

The densification of the accelerations for Z component with respect E and N components, between about $5.0 < t < 10.0$ s, could be correlated to the fact that in general the primary waves (p-waves) have lower periods than shear waves (s-wave), arriving at the surface first and vibrating parallel to Z. Also, by considering a certain point, vertically the soil could have a low stiffness thus it vibrates in accordance with the p-waves, whereas horizontally the soil opposes a great stiffness. Regarding the amplitudes, the s-waves provide values higher than p-waves (discussed in Chap. 1).

Figure 3.2 shows the shake map in terms of PGAs of the selected event. Usually, the PGA values refer only to the components E and N in the plane. It is possible to see the propagation of the PGA values, which reduce in function of the distance from the epicentre. The propagation is not perfectly circular and symmetric due to the soil characteristics that are different in each point altering this propagation. Graphically, this representation is like a "stone thrown into a lake" where the waves and energy propagate from a single point outward, which in this case is the epicentre.

In Fig. 3.2 they are indicated some general information of the earthquake (some of which already mentioned), whereas below there are some common symbols, as for instance, a star to indicate the epicentre position, and a triangle to indicate the seismic instrument, i.e., the station position (for more details see [9]).

Another important parameter of a time-history representation is the duration of the earthquake. It is defined as the time interval of the accelerometric signal in which the seismic motion is "significant". Two definitions are often used: (i) duration based on the exceedance of a threshold value called "bracketed duration". Here, a threshold is fixed, typically 0.05g, above which it is deemed that the motion has relevance for engineering purposes; thus, the duration is the time interval between the first and the last exceedance of this value; (ii) duration based on the motion intensity.

---

[3] With respect Appendix A in Chap. 8, this Butterworth filter, $B(\overline{\omega})$, adjusts the Fourier spectrum for the ground motion in frequency-domain, $\ddot{V}_g(i\overline{\omega})$, and by using FFT function the ground acceleration in time-domain, $\ddot{v}_g(t)$, is obtained.

For this, the Arias intensity function in time, $I_A(t)$, is calculated and normalized with respect to its maximum value, $I_{A,max}$; this $I_A(t)/I_{A,max}$ ratio is valid between 5.0 and 95.0%. The Arias intensity function is defined by:

$$I_A(t) = \frac{\pi}{2g} \int_0^t a^2(t)\mathrm{d}t \qquad (3.4)$$

where a(t) is the acceleration described in Fig. 3.1a, and g the gravity acceleration.

There is other intensity function, called Housner intensity, $I_H(\xi, T)$, defined by:

$$I_H(\xi, T) = \int_{0.1}^{2.5} S_{pv}(\xi, T)\mathrm{d}T \qquad (3.5)$$

where $S_{pv}(\xi, T)$ is the pseudo-velocity spectral response in function of the structural period, T, and the damping ratio, ξ (discussed in Chap. 2). Here, Eq. (3.5) is calculated with respect the range $0.1 \leq T \leq 2.5$ s, which includes building, civil and industrial structures, but it could be different.

The Arias intensity is related to the energy (physically represented by a "velocity") from the considered earthquake, whereas the Housner intensity is related to its potential damage ("displacement").[4] For the considered earthquake in E orientation, $I_{A,max} = 139.44$ cm/s, $I_H = 230.61$ cm, and the bracketed duration is 11.30 s.

To adopt this direct method of the time-history representation for estimating the seismic actions, at least 3 records of the same component are usually required to ensure that the PGA of all records do not fall below the PGA provided by code. In any case, it is possible to scale the values by using procedure called "ground motion scaling".[5]

When a spatial model of the structure is required, the seismic motion consists of three simultaneously acting accelerograms (i.e., E, N, Z component). The same accelerogram may not be used simultaneously along both horizontal directions to not overdesign the structure. For structures with special characteristics such that the assumption of the same excitation at all support points cannot reasonably be made, spatial models of the seismic action shall be used. Such spatial models shall be consistent with the elastic response spectra used for the basic definition of the seismic action.

---

[4] Note that the Arias intensity is directly correlated to the Fourier spectrum, which is also expressed as a velocity (see Appendix A in Chap. 8) through the earthquake energy; whereas Housner intensity, expressed in displacement, quantifies the damage which, for the building design, is correlated to the inter-storey displacement [14].

[5] The ground motion scaling consists in reducing or increasing a time-history representation in terms of displacements, velocities and accelerations to adjust it to specific needs for the design. However, it can be also used to adjust an elastic spectrum, the Arias intensity, etc. (for more details see [30]).

This method has the advantage that a structure can be directly analysed in a dynamic way for natural motions where all information is intrinsically considered (i.e., soil characteristics, amplification effects, etc.) [16, 31]. However, it is difficult to separate each information to understand its real effect and weight.

3.2.2 Elastic Spectrum from Time-History Representation

From time-history representation it is possible to obtain in a numerical way a specific elastic response spectrum. Each accelerogram in the t-domain provides an elastic spectrum in T-domain. The steps to carry out this transformation are [10]:

(i)   Downloading the values of the selected accelerogram in time-domain from a specific database. Similar for Fig. 3.1, the specific parameters for an earthquake (magnitude, distance, depth, date, etc.) must be chosen in order to obtain the relative histories.

(ii) Defining the damping ratios, ξ. The standard value is 5.0% however, it is possible to plot the elastic response spectrum for different ξ values (e.g., from 2.0 to 10.0%). Listing the natural structural periods, T, in small sampling interval.

(iii) Calculate the displacements of the structure in t-domain, by using the Duhamel integral described in Chap. 2. Find their maximum values of the displacements (called SD, in Fig. 3.3a) for each T, and their pseudo spectrum accelerations (PSAs) proportional to ω^2 ($= 4\pi^2/T^2$). Repeating these steps for all selected ξ values.

Figure 3.3 shows these elastic spectra relative to already selected Turkey 1999-year earthquake in terms of maximum displacements (Fig. 3.3a) and accelerations (Fig. 3.3b) for $\xi = 5.0\%$.

It is important to remember that the values of SD and PSA spectra provide intrinsically information about the earthquake and the structure, which is idealized as simple oscillator with a single-degree-of-freedom (SDOF). The PSA values for T = 0 correspond to the PGA peaks shown in Fig. 3.1a, thus in these cases, only the information about the earthquake is provided.

As expected, in the SD curves it is possible to see that a flexible structure (with high T values) generates higher displacements, in fact the curves tend to increase. The PSA shape for a real registration is obviously irregular in contrast to the elastic response spectrum by code (see Chap. 2). This is because the code must standardize a unique behaviour valid throughout a specific country, however some analogies are visible: an initial ascending branch up to high values, some peaks that maintain for a certain T duration (i.e., the constant branch in code), and a final descending branch. These trends are the main reasons of the shape adopted in the codes.

The considerations regarding the distance, soil and hazard effects, valid in a very general way for an elastic spectrum by codes, should be still valid for a real PSA however they are very difficult to individuate them. Also, the code elastic spectrum

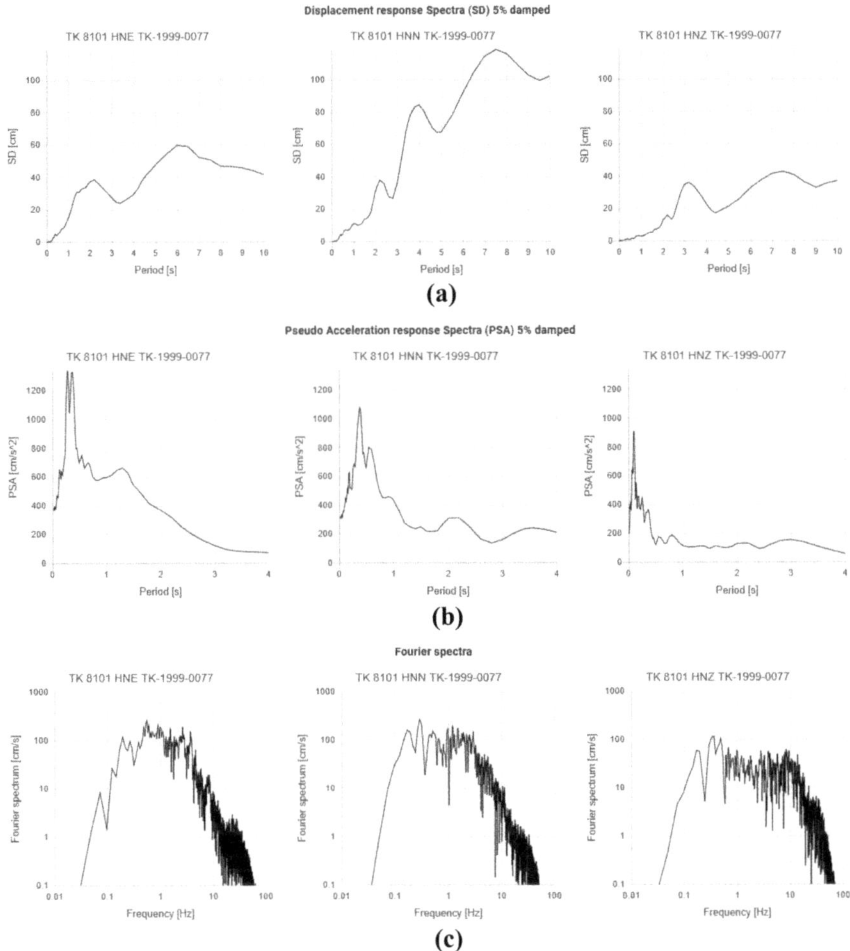

Fig. 3.3   Elastic spectra relative to Turkey 1999-year earthquake. Results in terms of **a** displacement and **b** accelerations for $\xi = 5.0\%$; **c** Fourier spectra (retrieved from ESM database [11])

is calibrated by considering several generic earthquakes, whereas Fig. 3.3 considers only one.

Figure 3.3c shows the Fourier spectrum for the ground motion of the 3 components generated by the considered earthquake (in Appendix A in Chap. 8 is explained the mathematical relation between the ground motion and the Fourier spectrum).

These spectra are very irregular and in general strongly decrease for high frequency ($f > 10.0$ Hz) where they appear denser. This trend decreasing, thus an energy decreasing, is mainly correlated to the already mentioned dissipation phenomena (e.g., frictions, dispersions, etc.). The utility of a spectrum is to know the amount of energy of an earthquake, and its peak values in order to avoid the resonance amplification that occurs when $f_{structure} \approx f_{earthquake}$.

Unlike SD and PSA spectra, a Fourier spectrum is plotted in f-domain, which represents the frequency of a harmonic forcing function (in this case the earthquake), thus this spectrum only provides information about the earthquake. Here there is no information regarding the structure.

Chapter 4
Artificial Accelerogram

Abstract In seismic regions where only limited data and records are available, generating artificial accelerograms can be a practical and effective solution for calculating seismic actions. Artificial accelerograms are designed to match the elastic response spectra required by codes, under restrictions consistent with the earthquake's magnitude and other relevant features influencing ground acceleration. Their generation is based on stochastic processes explained in this chapter as well as the spectral functions and possible adjustments.

4.1 General

In the seismic area where there are not many available data, parameters, registrations, etc. to calculate the seismic actions, the generation of an artificial accelerogram is a practical and useful option.

Artificial accelerograms shall be generated to match the elastic response spectra for code under 5.0% viscous damping. The duration of the accelerograms shall be consistent with the magnitude, and the other relevant features of the seismic event underlying the establishment of the ground acceleration. When site-specific data are not available, the duration of the stationary part of the accelerograms, T_s, should be $T_s > 10.0$ s.

The suite of artificial accelerograms should observe the following rules [14]:

(i) A minimum of 3 accelerograms should be used. Note that this condition has been also imposed for the direct method "time-history representation" (Sect. 3.2.1), in fact both for real earthquakes and for artificial earthquakes, the final output is substantially analogue, i.e., it is an accelerogram in t-domain.

(ii) The mean of the zero-period spectral response acceleration values (calculated from the individual time histories) should not be smaller than the value of the ground acceleration for the site in question. This rule would avoid underestimating the artificial peak ground acceleration.

(iii) In the range of periods, T_1, between 0.20 and 2.0 s, no value of the mean 5.0% damping elastic spectrum calculated from all-time histories, should be less

E. Zacchei and R. M.L.R.F. Brasil, *Calculation of Seismic Actions for Structural Analysis*, https://doi.org/10.1007/978-3-032-08884-0_4

than 90.0% of the corresponding value of the 5.0% damping elastic response spectrum. The period T_1 is the fundamental period of the structure in the direction where the accelerogram will be applied (this period as well as the sequent periods can be estimated by the well-known modal analyses [15]). This rule also avoids underestimating maximum spectral values.

The generation of an artificial accelerogram is carried out by a stochastic process explained in the next section.

4.2   Stationary and Ergodic Processes

A random process is a family of several random parameters related to a similar phenomenon, which may be functions of one or more independent variables [38]. Any sample waveform, involving one independent variable in t-domain taken from a real stationary process having a zero-mean value, can be separated into its frequency components using a standard Fourier analysis.

An earthquake artificial acceleration, a(t), can be defined as [15, 33]:

$$a(t) = I(t) \sum_{i=1}^{n} A_i \sin(\omega_i t + \phi_i) \tag{4.1}$$

where A_i is the fixed amplitude for each harmonic waveform (with the index of summation, i = 1, 2, …, n, where n is the harmonic waves number), ω_i is the circular frequency of the waves, and ϕ_i is the random phase angle between 0 and 2π. The latter parameter provides irregular and random acceleration functions.

As discussed, the Fourier series approximates a real function thus a high number of harmonics, n, would produce a better result. For this purpose, in a practical use, an order of 1500.0 harmonics (i.e., n = 1500.0) is usually sufficient, and it is not computationally heavy [16].

The function I(t) is the modulation function, which simulates the transitory behaviour of an earthquake. This function alters substantially the a(t) form to produce an artificial accelerogram like a real one. In this sense, it must be pre-defined in a deterministic way choosing an adequate form among those shown in Fig. 4.1, with the following relations, from Fig. 4.1a–d, respectively:

$$Rectangular\ function : I_a(t) = 1, \quad 0 \le t \le T_s \tag{4.2}$$

$$Trapezoidal\ function : I_b(t) = \begin{cases} \frac{t}{t_1}, & 0 \le t \le t_1 \\ 1, & t_1 \le t \le t_2 \\ \frac{t-t_3}{t_2-t_3}, & t_2 \le t \le t_3 \end{cases} \tag{4.3}$$

$$Exponential\ function : I_c(t) = c_1\left(e^{-c_2 t} - e^{-c_3 t}\right), \quad t \ge 0 \tag{4.4}$$

$$Semi-exponential\ function: I_d(t) = \begin{cases} \left(\frac{t}{t_1}\right)^2, & 0 \le t \le t_1 \\ 1, & t_1 \le t \le t_2 \\ e^{-c_1(t-t_2)}, & t \ge t_2 \end{cases} \qquad (4.5)$$

where c_1, c_2, c_3 are involved constants to be assigned after considering such factors as earthquake magnitude and epicentral distance, and t_1, t_2, t_3 are different time points. Obviously, in literature it is possible to find other modulation functions (e.g., logarithmic functions, triangular, etc.). Note that the maximum value is 1.0 since its purpose is to modulate the accelerations without altering their values.

In general way, Fig. 4.1b and Fig. 4.1d could be more consistent to a real accelerogram whit an ascending branch from zero to high values (beginning of the earthquake), a constant branch where high peaks of the accelerogram repeat in a certain time interval, and then a descending branch up to zero values (end of the earthquake). The trapezoidal modulation function is easier to describe mathematically than the others since it is composed by three linear functions.

Figure 4.1c also can be represent well an accelerogram trend, whereas Fig. 4.1a called "rectangular impulse" could be used for simulating other phenomenon; in fact, this model, in general, can be used for other purpose different to earthquake engineering.

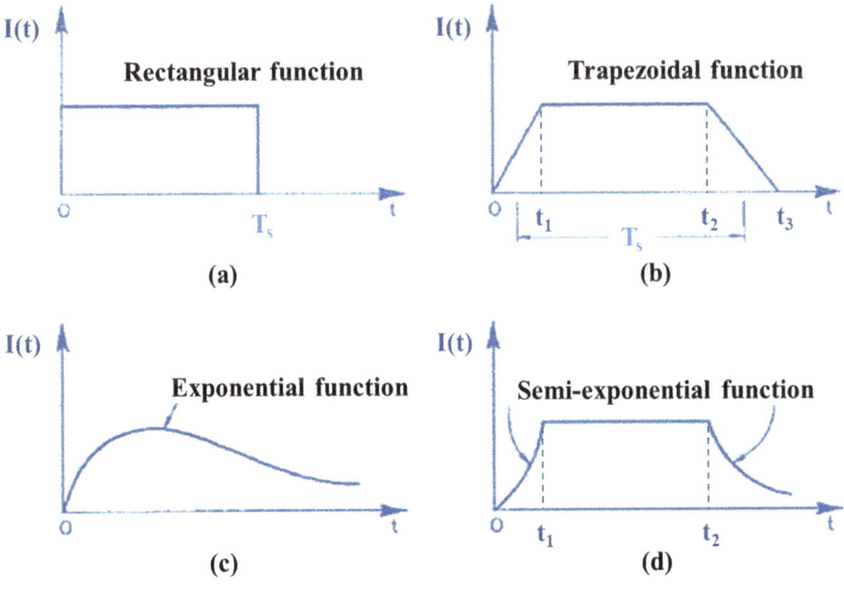

Fig. 4.1 Possible modulation functions (adapted from [33])

It is known that if the stationary process is ergodic,[1] the power spectral density (PSD) function, here defined by $G_z(\omega)$, is estimated as the Fourier transform of an autocorrelation function, $R_z(\tau)$, of any sample waveforms, $z(t)$:

$$G_z(\omega) = \frac{1}{2\pi} \int_{-\infty}^{\infty} R_z(\tau) e^{-i\omega t} d\tau \tag{4.6}$$

$$R_z(\tau) = \int_{-\infty}^{\infty} G_z(\omega) e^{i\omega t} d\omega \tag{4.7}$$

where i is the imaginary part, and τ also indicates the time, but it corresponds to a very short time like an impulse. Equations (4.6)–(4.7) are correlated each other by the Fourier integrals.

A PSD function provides information about the distribution of the power in function of the circular frequency, ω, of the structure (rigorously the power is quantified as an amount of the energy per unit time).

Note that the circular frequency in Eq. (4.1) refers to the earthquake, whereas in Eq. (4.6) refers to the structure (as shown later). Rigorously these frequencies are different however it is convenient to use the same symbol. This aspect introduces the concept of finding a PSD compatible with an elastic spectrum, since the frequencies that define the limits of the branches of a PSD are the same that delimit an elastic spectrum.

The relation between the energy and a structure has been already introduced by the force versus displacement in a simple oscillator, by for instance the energy dissipation using damping ratio, ξ (see Chap. 2). In this sense, a PSD correlates in a more direct way the energy with a structural system.

The $R_z(\tau)$ function (Eq. (4.7)) can be expressed in a most basic form by the covariance function:

$$R_z(\tau) \equiv E[z(t)z(t+\tau)], \tag{4.8}$$

which, like all ensemble averages, E, is independent of time t thus it is a function of τ only.

The variance of z(t), σ_z^2, can be described in a discrete system by:

$$\sigma_z^2 = \sum_{i=1}^{n} G_z(\omega_i) \Delta\omega_i = \sum_{i=1}^{n} \frac{A_i^2}{2} \tag{4.9}$$

---

[1] A stationary process is verified when all ensemble averages (i.e., mean value, variance, covariance, etc.) are independent of time t. The process is also ergodic when an ensemble average is equivalent to any function of a member of the ensemble in the time average. Therefore, an ergodic process must always be stationary, whereas a stationary process may or may not be ergodic [15].

from where it is possible to obtain the amplitude of each waveform, A_i, by the following approximation:

$$A_i \approx \sqrt{2G_z(\omega_i)\Delta\omega_i}, \tag{4.10}$$

thus Eq. (4.1) can be written in a practical way as:

$$a(t) = I(t) \sum_{i=1}^{n} \left(\sqrt{2G_z(\omega_i)\Delta\omega_i}\right) \sin(\omega_i t + \phi_i). \tag{4.11}$$

Therefore, to generate an artificial accelerogram, a(t), basically it is necessary to define only the modulation function, I(t), and the PSD function, $G_z(\omega)$.

4.2.1 Power Spectral Density Function Compatible with an Elastic Spectrum

Here the model proposed in literature [34] to define the PSD function has been described and explained. This model appears easier to be implemented and it provides analytical relations of PSD functions for elastic spectra with two, three, and four branches indicating thus a good versatility of the model. Figure 4.2 shows a PSD function compatible with an elastic response spectrum with four branches, which would represent the more complete elastic spectrum.

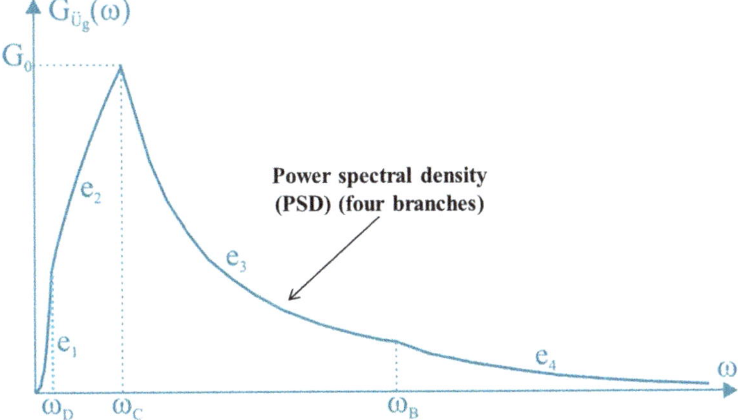

Fig. 4.2 Power spectral density (PSD) function with four branches (adapted from [34])

In accordance with [34, 37], $G_{\ddot{U}_g}(\omega)$ is the PSD function described by[2]:

$$G_{\ddot{U}_g}(\omega) = G_0 \left(\frac{\omega_D}{\omega_C}\right)^{e_2} \left(\frac{\omega}{\omega_D}\right)^{e_1}, \quad 0 \le \omega \le \omega_D \tag{4.12}$$

$$G_{\ddot{U}_g}(\omega) = G_0 \left(\frac{\omega}{\omega_C}\right)^{e_2}, \quad \omega_D \le \omega \le \omega_C \tag{4.13}$$

$$G_{\ddot{U}_g}(\omega) = G_0 \left(\frac{\omega}{\omega_C}\right)^{e_3}, \quad \omega_C \le \omega \le \omega_B \tag{4.14}$$

$$G_{\ddot{U}_g}(\omega) = G_0 \left(\frac{\omega_B}{\omega_C}\right)^{e_3} \left(\frac{\omega}{\omega_B}\right)^{e_4}, \quad \omega \ge \omega_B \tag{4.15}$$

where ω_B, ω_C, ω_D are the circular frequencies of the structure that limit each branch, the peak value of the PSD, G_0, is defined by:

$$G_0 = \frac{\gamma}{\beta_2 \omega_C} \left(\frac{\alpha S_0}{\eta_U^2 \omega_C}\right)^2, \quad with \ \gamma = \frac{4\xi}{(\pi - 4\xi)} \tag{4.16}$$

where ξ is the damping ratio of the structure, α is the dynamic amplification factor, and S_0 corresponds to the peak ground acceleration.

The meaning of these parameters is the same explained in Chap. 2 as well as the meaning of the ω_B, ω_C, ω_D points. The S_0 value is the maximum acceleration registered in an accelerogram, whereas the factor α regards the amplification of the accelerations mainly due to the site effects. The exponents e_1, e_2, e_3, e_4, and coefficient, β_2, substantially provide the branch shapes of the PSD functions. Their relations are described in Appendix B as well as the relation to define the correlated elastic response spectrum.

Another analogy with respect the elastic response spectrum regards the proportionally with ω frequency verified also for the PSD function, i.e., $G_{\ddot{U}_g}(\omega) = \omega^4 G_{U_g}(\omega)$.

Equation (4.16), and in general this model, represent the analytical closed forms to plot $G_{\ddot{U}_g}(\omega)$ curves, which can be obtained in an easy way by using parametric software. Thus, substantially, the model proposed in [34] regards Eqs. (4.12)–(4.15).

The factor η_U is called peak factor, which can be defined by:

$$\eta_U(\xi, \omega) \approx \sqrt{2 In\left[2N_U(\omega)\left(1 - e^{-\delta_U^{1.2}(\xi)\sqrt{\pi In[2N_U(\omega)]}}\right)\right]} \tag{4.17}$$

with

---

[2] The PSD function of any sample waveform, $z(t)$, becomes here a PSD for a specific wave, since, as discussed in Chap. 2, the elastic response spectrum is based on the shear waves that generate horizontal accelerations of the soil, here called $\ddot{U}_g(t)$; thus to maintain the same nomenclature used in [34] it is assumed that $G_z(\omega) \equiv G_{\ddot{U}_g}(\omega)$.

$$N_U(\omega) = -\frac{T_s \omega}{2\pi \, ln(p)} \tag{4.18}$$

$$\delta_U(\xi) = \sqrt{1 - \frac{1}{1-\xi^2}\left[1 - \frac{2}{\pi}\arctan\left(\frac{\xi}{\sqrt{1-\xi^2}}\right)\right]^2} \tag{4.19}$$

where T_s is the duration of the stationary part of the accelerograms (already mentioned), and p is the probability density function (or Gaussian distribution) of the peak factor η_U.

Note that Eq. (4.17) is an approximation obtained by the "first-passage problem"[3] for a linear system with a single-degree-of-freedom (SDOF) under a stationary process. It is identical to the spectral relative displacement, $S_d(\xi, \omega)$, thus here $S_{pa}(\xi, \omega) = \omega^2 \, S_d(\xi, \omega) \equiv \omega^2 \eta_U \sigma_U$ (see Chap. 2).

Given that this method is stochastic, a standard deviation, σ_U, which accounts for the possible error of the displacement values is introduced, as well as the spread factor $\delta_U(\xi)$, which here is physically correlated to the damping ratio. In [33], the factor δ_U is correlated to a fictitious damping ratio that can be assumed equal to the real one.

As mentioned, Eqs. (4.12)–(4.15) are also valid for elastic spectrum with different number of branches, therefore with $\omega_D \to 0$ the elastic spectrum has three-branches, and with $\omega_D \to 0$, $\omega_B \to \infty$ it has two-branches.

4.2.2 Adjustments of the Artificial Acceleration

The obtained artificial acceleration by Eq. (4.11) derivates from stochastic processes and under some theoretical approximations. Thus, even if the elastic response spectrum is correctly transformed to an artificial accelerogram by PSD function, the algorithm could provide some incongruences. In fact, the main goal of the used analytical model is to define a PSD function coherent with the elastic spectrum, therefore, the main aspect to be respected is the power distribution (or energy content) versus the circular frequency ω.

One incongruence regards the displacement values that must be null at $t = 0$, which can be solved by using the tool "baseline correction" (already described in Chap. 3). Another incongruence regards the maximum value acceleration since it must be the same in the artificial accelerogram and in the elastic response spectrum (i.e., $|a(t)| = S_0$). This can be corrected in two easy ways:

---

[3] The first-passage problem for random processes can be defined as the determination of the time (here in ω domain) taken for a variable to reach a certain peak. Due to the lack of a closed-form solution under general conditions, several approximations have been derived (for more details see [36]).

(i) If it is verified that $|a(t)| < S_0$, only the maximum $|a(t)|$ value must be amplified up to S_0 value. This correction would allow that only one peak is shown in the artificial accelerogram, which in an elastic spectrum must be unique.

(ii) If $|a(t)| > S_0$, all maximum $|a(t)|$ values must be reduced in a proportional way in order to obtain a unique S_0 value.

In general, in analogue way to a real accelerogram also for an artificial accelerogram it is possible to scale all values to generate more specific scenarios and, for instance, to consider amplification phenomena.

Figure 4.3 shows an example of how the results can be represented: from an elastic response spectrum, a PSD is defined and then, an artificial accelerogram is produced. Non-adjusted and adjusted accelerogram is also shown by black and grey curves; the latter maintains the peaks under the S_0 value provided by the elastic spectrum (in this example $S_0 = 0.34g$, indicated also by the horizontal dashed lines in Fig. 4.3c). This analytical method can be solved easily by using adequate software (e.g., Wolfram Mathematica software, etc.). It is possible to see the effects of the modulation of $I(t)$ function for a $T_s = 11.0$ s (here adopted Fig. 4.1b by Eq. (4.3)).

Note that, this method generates an accelerogram denser that a real accelerogram, this is because the whole energy of the elastic spectrum passes entirely to the artificial accelerogram without any dissipation. In fact, as mentioned the elastic spectrum by code overestimates its values thus the energy involved in the system.

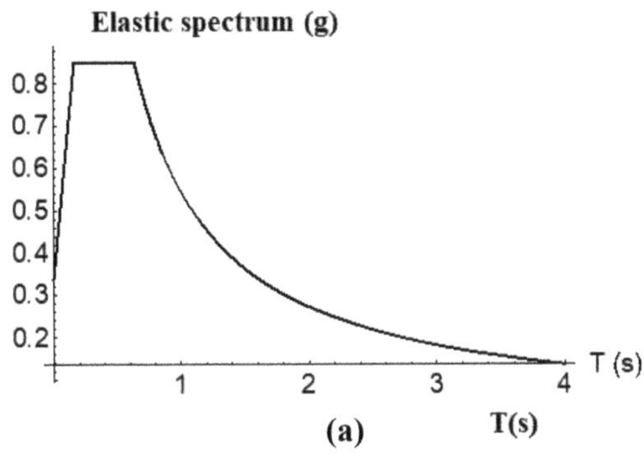

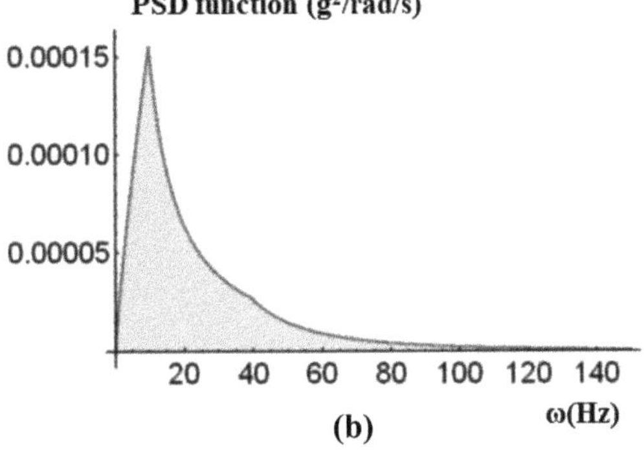

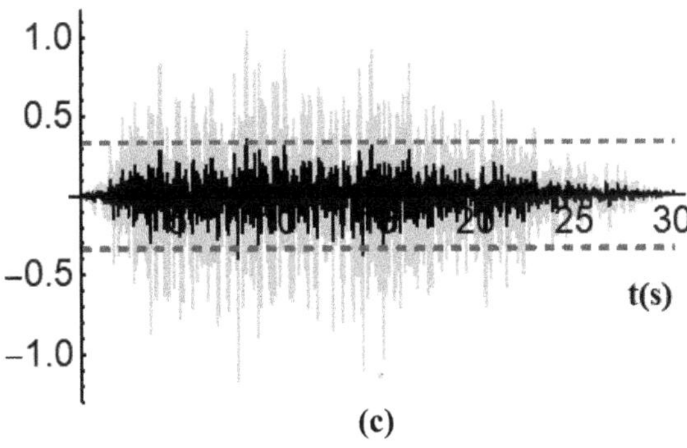

Fig. 4.3 Example of the model application: **a** elastic spectrum; **b** PSD function; **c** artificial accelerograms (modified by [35])

Chapter 5
Seismic Hazard Analyses

Abstract The seismic hazard analyses are considered the best methods to estimate seismic inputs, since they identify and combine, in both mathematical and physical ways, many uncertainties involved in a seismic event. They account for the seismic-geological context and historical earthquakes through seismogenic zones and attenuation equations. The analyses can be developed in probabilistic or deterministic ways, providing several scenarios. The former provides the most likely earthquake, whereas the latter provides the worst-case earthquake.

5.1 Probabilistic Approach

The probabilistic seismic hazard analysis (PSHA) is considered the better method to estimate the spectral accelerations, since it identifies, quantifies and combines in mathematical and physical way, the uncertainties involved in a seismic event, e.g., the size, location, rate of occurrence of the earthquakes, the variation of ground motion characteristics, etc.

The goal of this method is to provide the most likely earthquake. Given that it is based on seismic hazard maps, it could be considered as a "spatial model" to calculate the seismic actions.

5.1.1 Seismogenic Zone and Model

A seismogenic zone (ZS) is conceptualized through a combination of geological and geophysical information. The boundaries of these seismogenic zones are defined in function of the collected data of the historical earthquakes, local seismicity, tectonic configuration, location of active faults, etc.

These zones are defined by polygons where it is assumed that the seismicity is uniform in terms of earthquake types and distributions. It is meant as "earthquake type", the earthquake generated by the same failure mechanism

E. Zacchei and R. M.L.R.F. Brasil, *Calculation of Seismic Actions for Structural Analysis*, https://doi.org/10.1007/978-3-032-08884-0_5

(discussed in Chap. 1), and as "earthquake distribution" the same probably of occurrence in a specific zone.

In fact, the main hypotheses of a seismogenic zone are that:

(i)   The earthquakes are equally likely in space. This indicates that each earthquake in each point of the ZS can occur with the same probability.

(ii) There is the same occurrence (or frequency) of each earthquake.[1] This indicates that in a ZS the rate of occurrence of the earthquakes is the same, and it maintains constant. This can be considered as a likely in time.

(iii) For each zone it is defined a unique predominant failure mechanism. Despite the earth is characterized by several failures and cracks, the definition of a unique a simple mechanism is useful for mathematical and physical models.

In general, these hypotheses can be reasonably accepted if it is considered that, in the very long term (e.g., 100,000.0 years), the deformations of the block (i.e., seismogenic zone times depth) are linear elastic, and they maintain within the block not affecting nearby blocks. These zones are usually calibrated for a depth < 30.0 km where only superficial earthquakes are generated.

It is possible to find maps indicating seismogenic zones of different countries in the world by literature, databases, websites, etc. (among others [6, 43, 46]). A very good example is the map of the seismogenic zones for the Iberian Peninsula called "ZESIS" database [42] as shown in Fig. 5.1. This is an interactive database where it is possible to obtain general information and key parameters to carry out the PSHA analysis.

Figure 5.1a shows the 55 superficial seismogenic zones and the faults (about 200 faults represented by irregular lines), whereas Fig. 5.1b shows the earthquakes (points).

Where there is a high concentration of earthquakes, there are also a high concentration of active seismic faults thus the ZSs are designed by a smaller area in order to distinguish well the main failure mechanism for each earthquake group. Obviously, the discretizing of the ZSs is direct correlate to the available data.

The formulation of the PSHA analysis is based on Cornell model [40], where the mentioned hypotheses allow to develop it. This model is divided in four steps as shown in Fig. 5.2:

Step 1. Identification and characterization of all earthquake sources. The seismic sources can be represented by rectangular area, line, circular, etc. to develop analytically the model. However, by using software (e.g., Crisis software), it is possible to use sources with more irregular areas provided by the described maps of the seismogenic zones.

Step 2. Plotting of the recurrence relationship. The Gutenberg-Richter (G-R) law (Eq. (5.2)) is usually used for this purpose. This law correlates the recurrence of

---

[1] The ideal concept is that the occurrence of one event does not change the probability of occurrence of any other event. Obviously, this is a weak hypothesis since the deformations of the block due to an earthquake alter other potential crack points anticipating or postponing other earthquakes. However, for this model is usually used the Poisson distribution, thus the independency and stationarity among earthquakes must be maintained [38].

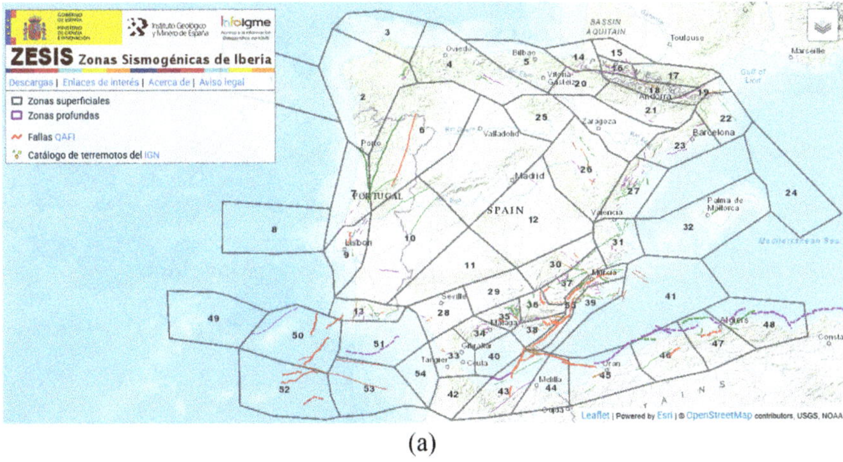

(a)

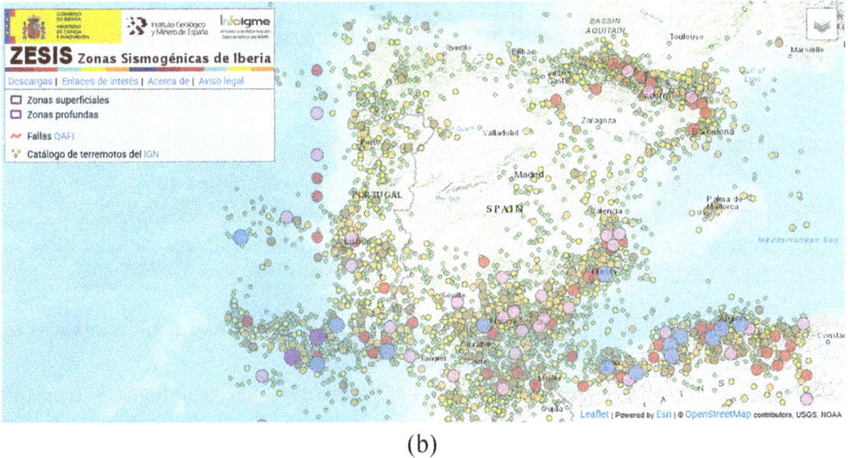

(b)

Fig. 5.1 Map for the Iberian Peninsula: **a** seismogenic zones and faults; **b** earthquakes (retrieved from [42]). The caption and other information are provided in Spanish, however they are also explained in the text

the earthquakes with respect a magnitude, to obtain the a-value and b-value.[2] This operation can be done by knowing the number of events for each magnitude and by defining the completeness period (discussed later). In this sense, it provides a probability distribution.

Step 3. Choosing of an attenuation relation. In literature there are several attenuation equations that describe the ground motion in function of the distance and

---

[2] Note that, by using ZESIS database this step can be bypassed since this database already provides these parameters (λ_c, a-value, b-values, etc.). However, in general for almost of countries in the world this step must be developed manually, using Microsoft Excel or other popular tools, by downloading the historical catalogue of the earthquakes.

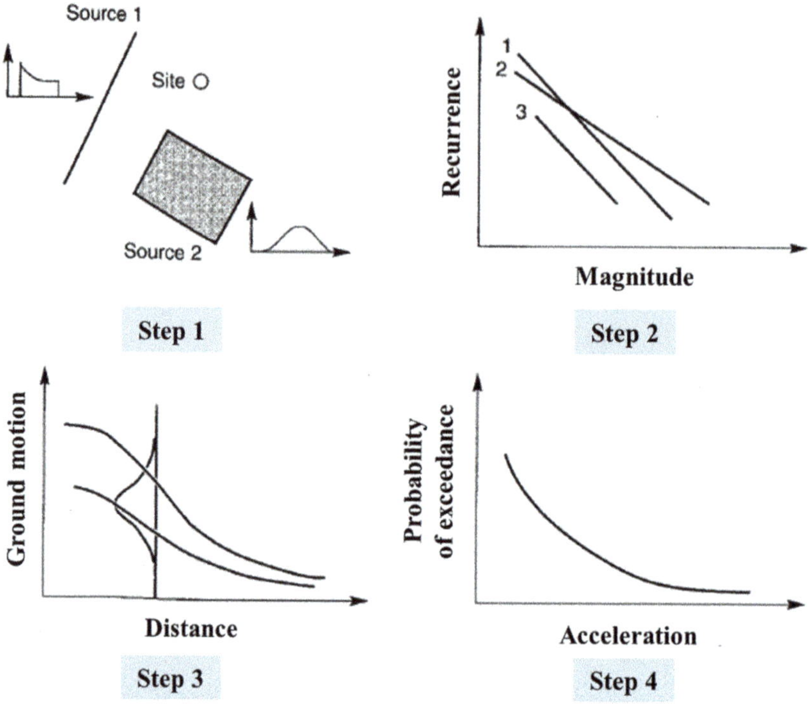

Fig. 5.2 Four steps for developing the PSHA analysis (adapted from [3])

magnitude. This relation plays an important role in the results of the analyses, for this a separated section is dedicated (shown later).

Step 4. Obtaining of the results in terms of probability of exceedance of a certain acceleration with respect other, elastic response spectra for different return period, disaggregation results, etc.

The probability, P, that a defined ground motion acceleration, Y, exceeds a possible peak ground acceleration, y^*, in an i-th seismogenic zone can be defined by:

$$\Lambda_i = \lambda_c \int\limits_m \int\limits_r \int\limits_\sigma P\big[Y > y^* | m, r, \sigma\big] f_m(m) f_r(r) f_\sigma(\sigma) \, dm \, dr \, d\sigma \qquad (5.1)$$

where λ_c is the annual rate of the exceedance of a magnitude, m, with respect other one in a seismogenic zone, correlated by the recurrence G-R law:

$$\log(\lambda_c) = a - (bm) \qquad (5.2)$$

where the a-value is the intercept of this linear law, which indicates the level of seismicity. The higher the seismicity of the region, the greater the value of a (a reference high value could be a $\approx$ 3.0). The b-value is the slope of Eq. (5.2), which indicates the ratio between the number of small and large events. The value of b = 1.0 is a universal reference, with b < 1.0 the zone should be dominated by small and large earthquakes, whereas with b > 1.0 the zone should be dominated only by small earthquakes. These parameters are very important for the analyses; thus, they must be well estimated.

The functions $f_m(m)$ and $f_r(r)$ are the probability density function (PDF) of the magnitude, m, and of the site-source distance, r, respectively:

$$f_m(m) = \frac{\beta e^{\beta(m - M_{min})}}{1 - e^{\beta(M_{max} - M_{min})}} \tag{5.3}$$

$$f_r(r) = \frac{2r}{r_{max}^2 - r_{min}^2} \tag{5.4}$$

where M_{min} and M_{max} are the minimum and maximum magnitude, and r_{min} and r_{max} are the minimum and maximum site-source distance, respectively. The β-value is correlated to b-value by the following relation: $\beta = b(In10)$.

Both functions are used to specify the probability of the random variable (magnitude and distance) falling within a particular range of values, rather than taking on any value. Thus, they must be pre-defined in accordance with the physical model to be simulated. Obviously, there are different relationships in literature.

The $f_m(m)$ function is correlated to G-R law and Poisson distribution, whereas the $f_r(r)$ function is correlated to the geometry of the seismogenic zone (discussed for step 1). The $f_\sigma(\sigma)$ function is the Gaussian distribution, which accounts for the standard deviation, σ (i.e., statistical errors), of the ground motion randomness. This function is often not shown in Eq. (5.1) since it could be included directly in the attenuation equation.

Usually, the analysis involves several seismogenic zones, N, thus Eq. (5.1) must be considered the summation $\sum_{i=1}^{N} \Lambda_i$. To understand the model in the analytical way, the case with a unique seismogenic zone is considered (i.e., i = 1), thus the probability of exceedance of an earthquake is defined as:

$$\Lambda_1 \equiv \int_{r_{min}}^{r_{max}} P[M > M^*] \frac{2r}{r_{max}^2 - r_{min}^2} \, dr$$

$$= \int_{r_{min}}^{r_{max}} \frac{1}{1 - e^{-\beta(M_{max} - M_{min})}} \left\{ e^{-\beta(M^* - M_{min})} - e^{-\beta(M_{max} - M_{min})} \right\} \frac{2r}{r_{max}^2 - r_{min}^2} \, dr. \tag{5.5}$$

The magnitude M_{min} and M_{max} are defined in function of the estimated probability of exceedance for a certain M^*.[3]

---

[3] Note that Eq. (5.5) refers to the probability of exceedance of an earthquake considering its magnitude, m ($\equiv$ M, since a capital letter is usually used), and not its ground motion acceleration, Y, as shown in Eq. (5.1). In fact, Eq. (5.1) is written in a more general way.

The site-source distance refers to the $f_r(r)$ function, thus if the site where the PSHA is made coincides with the position of the structure, r_{min} can be assumed $r_{min} = 0$, whereas r_{max} can assume high values up to $r_{max} \approx 100.0$ km [44, 45].

It is important to highlight that r_{max} should be consistent to the size of the seismogenic zone and to the adopted attenuation equation. Therefore, it should correspond to a superficial distance (i.e., site-epicentre distance) that remains inside of the seismogenic zone, and it should assume values not greater than those used to calibrate the attenuation equation.

The M_{min} and M_{max}, referring to the $f_m(m)$ function, can assume different values in a seismogenic zone. In general, $M_{min} = 4.0$ (a value of $M_{min} < 4.0$ usually does not generate damages to the civil structures and buildings), whereas M_{max} depends on the registered events.

Finally, the probability of non-exceedance and, another important parameter as the return period, T_r, should be calculated. They can be estimated, for a certain peak ground acceleration, y^* (or certain magnitude, M^*), by, respectively:

$$P\left[Y \leq y^*\right] = e^{-\lambda_c P\left[Y > y^*\right]} \tag{5.6}$$

$$T_r = \frac{1}{1 - P\left[Y \leq y^*\right]}, \tag{5.7}$$

which T_r period usually assumes very high values of the order of 10^2–10^3 years. It can be difficult to understand for a common user, and in any case, they are often misleading values, however it is adopted as a "trick" to increase the level of attention of designers on the seismic design.

The outputs of interest of the PSHA analysis are:

(i) The uniform hazard spectra (UHSs). They provide the spectral accelerations in function of the structural period, T, by accounting for the contribution to hazard of small-near and strong-far earthquakes for a unique return period in a specific site. They are analogues to those described in Chap. 3 (i.e., PSA curves versus structural periods), with the difference that UHS curves do not correspond to the expected movement of a single earthquake, but they represent the result of several possible earthquakes with different magnitudes and distances occurring at different sites (Fig. 5.3). It is possible to plot the peak ground accelerations and spectral acceleration curves in function of the return period, T_r, in a specific site; also, in these cases the contribution to hazard of several ZSs have been considered in each curve.

(ii) The disaggregated results. These results come from the disaggregation analysis, where it is possible to quantify whether a strong-far earthquake is worse than a small-near one under a certain return period and certain acceleration in a specific site. This analysis defines the hazard contribution for two key parameters (i.e., distance and magnitude) in a separated way. This analysis provides some magnitude and distance pairs, which correspond to possible scenarios (Fig. 5.4).

Fig. 5.3 Example of UHS
curves (adapted from [44])

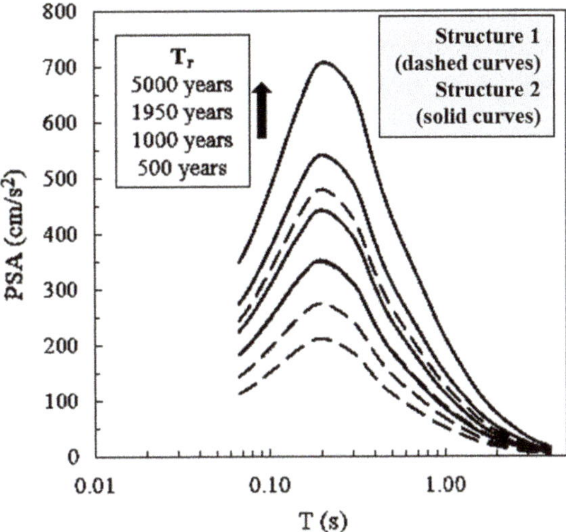

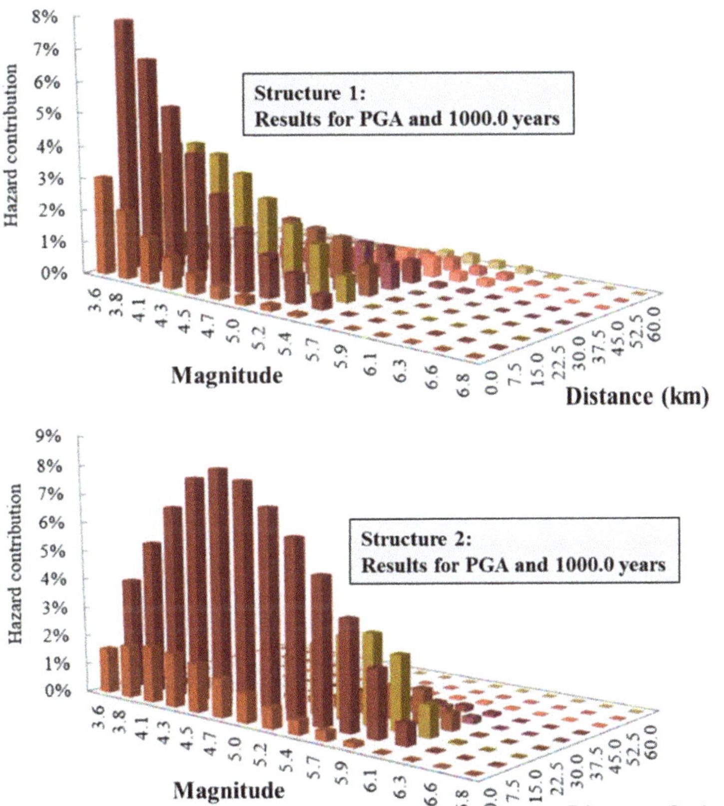

Fig. 5.4 Example of results by the disaggregation analysis (adapted from [44])

Figure 5.3 shows an example of the UHS curves where it is possible to see that the curves increase for high T_r periods.[4] Note that each curve maintains the same T_r period value along the T values.

Figure 5.4 shows an example of results by the disaggregation analysis calculated for a return period of 1000.0 years and for peak ground acceleration (PGA) values. Usually, they are represented by 3D histograms where the magnitude and distance are indicated. In this example, both structures could be subjected to a PGA of an earthquake up to 7.50 km away, with a low magnitude of 3.6 M (highest bar) for the structure 1, and a medium magnitude of 4.5 M (highest bar) for the structure 2. Therefore, the obtained pairs would be 3.6 M—7.5 km, and 4.5 M—7.5 km. Note that the hazard contribution is mathematically a PDF function.

To develop a PSHA analysis, they are necessary the seismic zones, earthquake catalogues (historical and/or instrumental), and attenuation equations. Also, adequate software could be necessary to solve Eq. (5.1) for N zones.

The PSHA being, as mentioned, a better analysis to calculate the seismic actions, it is possible to find pre-defined interactive maps as, for instance, the GEM database [47][5] where the seismic hazard is estimated in any points of the world.

However, this method has some criticisms to be discussed. There are several uncertainties called "aleatory" and "epistemic" uncertainties, where the former are related to the intrinsic randomness in the model described by PDFs, statistical errors, etc. thus they cannot be eliminated; whereas the latter are related to the lack of knowledge which could be difficult to be quantified since nowadays it is not known when and where the next earthquake will occur.

Also, the UHS curves do not account for the path and site effects in a satisfactory way although the path and site effects are included intrinsically in the attenuation relation. It is practically impossible to measure point-by-point the seismic waves and the material properties from a seismic source to a site. These effects are different for each location and for each earthquake at the same location, in fact each earthquake brings new data. Other criticism regards the used attenuation equations (discussed later).

---

[4] For very high T_r values, the PSHA analysis could lose its meaning and no longer be reliable. In the best of cases, it is possible to retrieve a good quantity of events with reliable information from the 18th century, thus it could be possible to define scenario for the next ~ 400.0 years covering a return period of 800.0 years. Therefore, for T_r > 1000.0 years, the PSHA appears not more valid, providing values, in this example, up to about 0.45g.

[5] This database provides interesting maps in the world. They provide not only the seismic hazard but also the seismic risk in general, which is calculated by the following relation: risk = hazard × vulnerable × exposure (mentioned in Chap. 2). They can be associated, to summarize the concept, with the following three words: hazard ↔ event; exposure ↔ losses; vulnerability ↔ structure [20].

5.1.2 Treatment of the Data

As mentioned, by using for example the ZESIS database [42], the key parameters
(λ_c, a-value, b-values, magnitude, etc.) needed to perform the PSHA analysis, are
already available and ready to be used. On the contrary, if these parameters are not
available (as usual), a pre- and post-treatment of the data is necessary.

The treatment of the data consists in three operations called: "homogenization"
of the magnitude, "declustering" of the events, and "completeness" of the catalogue
[49].

The homogenization consists in transforming all magnitudes to a unique type
of magnitude. This is because in the same seismic catalogue, it is possible to find
events registered by different magnitudes due to the issues discussed in Chap. 1. If it
is necessary to use different seismic catalogues the homogenization appears obvious.
In literature there are several correlations between different magnitudes, however it
is usually adopted the surface wave magnitude, M_s, and/or the moment magnitude,
M_w [48].

The declustering consists in eliminating the foreshocks and the aftershocks, thus
only the mainshocks are considered. This operation avoids "contaminating" the analyses by considering the same event more times providing a wrong weight regarding
its occurrence. As mentioned, the PSHA analysis is based on the Poisson process,
which implies that the events must be independent between each other. This second
operation also helps to reduce significantly the quantity of data to be treated.

The completeness analysis helps to find the period (in years) where the catalogue
contains all earthquakes that occurred under a certain magnitude. Thus, in a defined
completeness period the events have a regular frequency. This analysis substantially
would allow to estimate a correct λ_c value.

It is possible to find some algorithms to develop these treatments, however, a
"visual procedure" in many cases is sufficient by filtering the data by, for instance,
Microsoft Excel (for more details see [45, 49]).

5.1.3 Attenuation Equations

The attenuation equations, or ground motion prediction equations (GMPEs), need
a specific section. This is because they play an important role in the outputs of the
PSHA analyses, in fact they provide directly the spectral accelerations to be used for
design.

The choice of one equation instead of another mainly regards the geological,
geophysics, and geotechnical context, robustness of the data, considered distance
(i.e., epicentre-structure distance), etc. All these characteristics affect the calibration
of the attenuation equation. It would be ideal to adopt an attenuation equation calibrated in the same region where the structure is placed, otherwise, it is advisable to
adopt an attenuation equation calibrated in a similar seismic context.

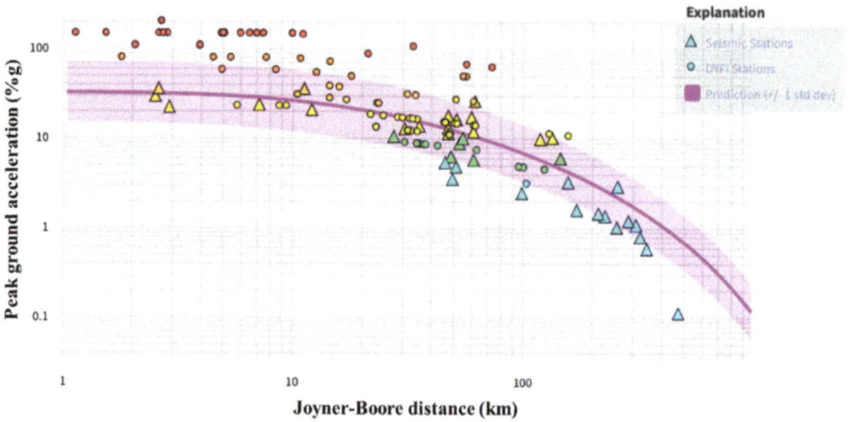

Fig. 5.5 Attenuation equation for soil (adapted from [9])

In [41] there are several attenuation equations (about 800 equations) for PGA and for spectral ordinates in the world. More in general, an attenuation equation can be defined as follows:

$$\log_{10} Y = \left[a_1 + a_2 m + a_3 \left(\log_{10} r \right) + \cdots + a_i F_i \right] \pm \sigma \tag{5.8}$$

where the coefficients a_1, a_2, ..., a_i, serve to adjust the relative parameters to be determined. The factors F_1, ..., F_i, could represent the site conditions (e.g., rock, soil), style of faulting, etc. The parameters Y, m (usually > 4.0), r (usually < 100.0 km), σ, have been already defined for Eq. (5.1).

Note that Eq. (5.8) is purely an example. Given that about 800 attenuation equations exist, it is obvious that several forms also exist. The two key parameters that are always present are the magnitude, m, and distance, r, which can be represented in different ways (discussed in Chap. 1), whereas other parameters are secondary but still important.

Figure 5.5 shows a possible logarithmic trend of an attenuation equation for soil retrieved from the database [9] where the PGA values are plotted in function of the Joyner-Boore distance (Chap. 1). The observations are represented by triangles that indicate the seismic stations and by points that indicate the "did you feel it" (DYFI) stations.[6] The prediction is represented by the solid curve between a band that varies of $\pm \sigma$.

The main criticisms of attenuation equations are correlated to the fact that they are often estimated by stochastic models, and they are calibrated for a unique specific region usually relatively small, thus they are subjected to provide wrong outputs when applied in different places.

---

[6] The DYFI system was developed to tap the abundant information available about earthquakes from the people who experience them.

5.2 Deterministic Approach

5.2.1 General

The deterministic seismic hazard analysis (DSHA) is like the previous PSHA analysis since it also considers the seismic-geological context and the historical earthquakes, but it appears easier. The goal is to provide the worst earthquake, thus it is usually carried to structures for which their failures could have catastrophic consequences, such as nuclear power plants and large dams.

For dams, for example, there are two main levels of earthquakes for structural analysis and design, i.e., the operating basis earthquake (OBE), and the safety evaluation earthquake (SEE) [50, 54]. The OBE level can be considered as an economical criterion and it is "negotiable", whereas the SEE level is a "non-negotiable" safety criterion. In this sense, SEE represents the maximum level of ground motion for which the dam should be designed or analysed, thus a DSHA is more appropriate to be developed by using up to a return period of 10,000.0 years; whereas OBE represents the level of ground motion at the dam site for which only minor damage is acceptable, thus it can be developed a PSHA analysis, and in many cases, it will be appropriate to choose a minimum return period of 145.0 years.

5.2.2 Model and Analysis

The main hypothesis of the model is that the return period, T_r, is so long that the PSHA analysis is considered unreliable [44, 45]. This time is estimated ideally $T_r > 10,000.0$ years, however, as already mentioned for PSHA analysis, from $T_r > 1000.0$ years a DSHA analysis should be used.

For the DSHA analysis, some steps defined for PSHA are still valid (see Fig. 5.2) [3, 51]. The application is deterministic, however the fact that the seismic hazard is considered, rigorously it can be considered as a semi-deterministic analysis.

The main steps are explained below:

Step 1. Identification and characterization of all earthquake sources (like step 1 for PSHA). Here the sources that can produce significant earthquakes must be identified, however, on the contrary of the PSHA, here only the values of the magnitude (M_1, M_2, ..., M_i) are necessary.

Step 2. Choosing of an attenuation relation (like step 3 for PSHA). Here the magnitude-distance pair to be used to plot the attenuation equation define the "controlling earthquake", i.e., the earthquake that is expecting to produce the strongest level of shaking. The magnitude can be retrieved from step 1, whereas the source-site distance must be defined a-priori (usually the shortest distance is selected).

Figure 5.6 shows the mentioned two steps to develop the DSHA analysis.

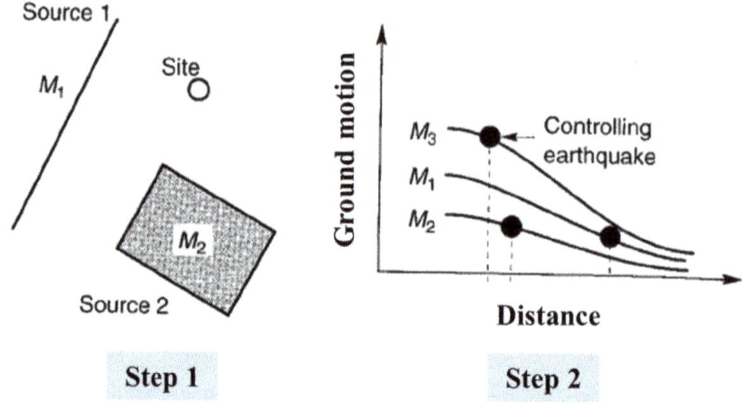

Fig. 5.6 Two steps for developing the DSHA analysis (adapted from [3])

This method is easy to be developed, in fact once defining some magnitude-distance pairs, those that produced the worst acceleration at the site can be identified directly as the "controlling earthquake". The main criticisms are that the DSHA analysis provides an estimate of ground motion without assessing the level of conservatism, and it does not account for the frequency of earthquake occurrence.

It is interesting to consider that this determinist method could be applied in three different ways:

(i) As mentioned, by selecting some high magnitude and short distances from different seismogenic zones. By using an attenuation equation (among others [52, 53]), the pair that produces the worst acceleration is the controlling earthquake. Here, the seismic hazard is considered since the knowledge of the seismogenic zones helps to define the significant earthquake produced by different failure mechanisms. This is the classical application of the DSHA analysis (Fig. 5.7 shows this application).

(ii) From the disaggregation analysis (see Fig. 5.4), a magnitude-distance pair is also obtained. From this pair it is possible to estimate the spectral accelerations by using an attenuation equation. In this case the approach is still deterministic, and it accounts for the seismic hazard, however it does not provide the worst scenario. In fact, the disaggregation analysis usually excludes the strong earthquakes, which are infrequent and occur over long periods of time (the completeness analysis tends to exclude strong and old earthquakes).

(iii) An attenuation equation can be used by applying any magnitude-distance par. Thus, it is possible to define ideal magnitudes and distances for estimating the spectra accelerations. This approach could be useful to have a real elastic response spectrum defined in a specific site accounting for directly the amplifications, site effects, etc. In this sense, the seismic hazard should be included intrinsically in the attenuation equation. By using this way, the obtained spectrum could be called "synthetic elastic response spectrum".

Fig. 5.7 Example of DSHA applications in terms of PSA curves (adapted form [45])

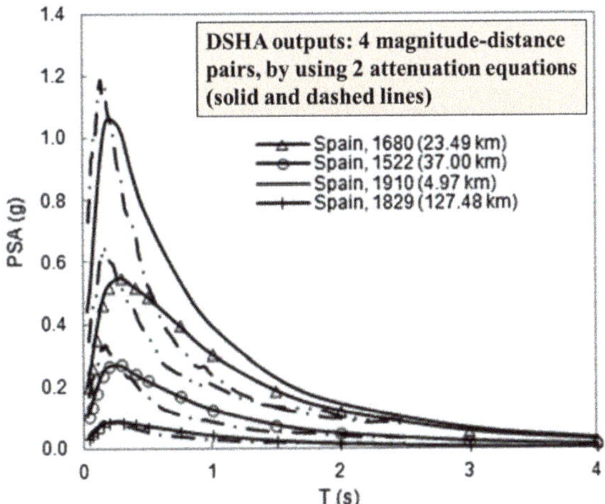

Figure 5.7 shows an example of pseudo spectrum accelerations (PSAs) in function of the structural period, T, obtained by the DSHA analysis. Here they are used two different attenuation equations for comparison, and four magnitude-distance pairs (they are indicated the country, year, and source-site distance).

In general, a near earthquake (4.97 km) generates high accelerations indicating that the distance can play a more important role with respect the magnitude, in fact the highest curves correspond to a magnitude of 6.1 M_w, whereas other ones correspond to values between 6.5 and 6.8 M_w (for more details see [45]). All curves decrease by increasing the distance. It is possible to see that DSHA curves can provide very high values up to about 1.20g, which is the worst acceleration.

Chapter 6
Physically Based Seismic Hazard Analysis

Abstract The physically based seismic hazard analysis is probably the most advanced analysis, computationally, compared to the others described in this book. For this reason, although this method also accounts for seismic hazard analysis like those explained in Chap. 5, it is described separately. This method follows the same procedures as the seismic hazard analyses, with only one main difference: the attenuation equations are not used, and they are replaced by empirical Green's functions to estimate the spatial soil displacement along the wave path.

The physically based seismic hazard analysis (pb-PSHA) is probably the most advanced analysis, in a computational way, with respect to the other ones described in this book. Thus, it is also the most complicate to be applied. For this reason, although this method also accounts for the seismic hazard analysis like those explained in Chap. 5, it is described separately.

This method follows the same procedures of the PSHA analysis with only one main difference, i.e., the attenuation equations are not used, and they are substituted in the model by empirical Green's functions (EGFs).

In accordance with code [14], accelerograms generated through a numerical simulation of source and travel path mechanisms, may be used, if the used samples are adequately qualified regarding the seismogenic features of the sources and to the soil conditions appropriate to the site, and their values are scaled to the value of ground acceleration for the zone under consideration.

These aspects would regard the model studied here where the outputs are simulated by accelerograms accounting for seismogenic characteristics, wave path effects, etc. However, this model is not yet consolidated and therefore is not found in codes. This model is also treated to estimate the artificial accelerograms non-compatible with an elastic response spectrum (not studied in this book; for more details see [33]).

6.1 Model

This model is based on the physical properties of the seismic source and of the elastic medium where the waves propagate. This model allows the direct estimations of the seismic parameters, as magnitude, focal distance, fault dimension, etc. However, some parameters are difficult to be well identified thus they represent the uncertainties into the computation of ground motion, unlike the classical PSHA developed under unbounded probability distributions.

The output from the pb-PSHA analysis is a series of source- and site-specific ground-motions that would comprise all earthquakes that could affect a site [55, 56, 58]. The steps to carry out the pb-PSHA analysis is shown in Fig. 6.1. Some steps of the pb-PSHA are like to the PSHA analysis, in particular the steps 1, 2 and 4, thus for brevity they are only recalled:

Step 1. Identification and characterization of all earthquake sources. Here, the fault rupture scenarios represent the source region.

Step 2. Plotting of the recurrence relationship by using the well-known Gutenberg-Richter (G-R) law. Steps 1–2 substantially serve to identify well the parameters to be used to define mathematically the seismic source.

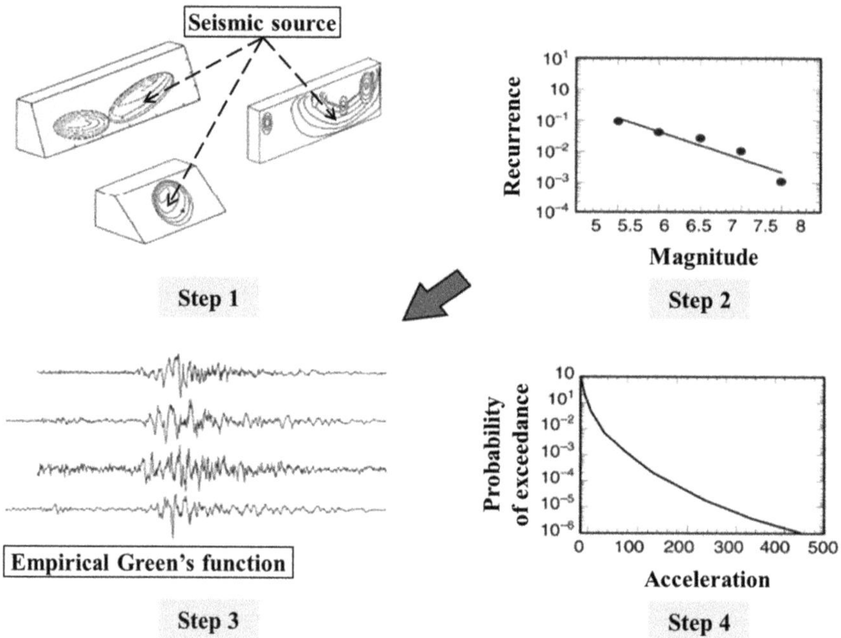

Fig. 6.1 Steps to develop the pb-PSHA analysis (adapted from [55])

Step 3. Developing of EG functions. This is the main step of this method where the synthesized[1] ground motions are obtained. The EGFs should replace the aleatory uncertainties that PSHA analysis estimate by using attenuation equations for specific sites [59]. The EGF function provides directly the soil motion in the plane in terms of displacements, velocities and accelerations, accounting for the properties of the seismic source. Its application is not easy, thus relative advanced computational analyses are necessary, as explained later.

Step 4. Probability of exceedance of a certain acceleration with respect other one. Seismic hazard curves for a particular synthesized ground motion are obtained.

Actually, the pb-PSHA follows the same steps of the standard PSHA analysis, but it replaces the attenuation equation with calculations of physically based synthetic seismograms, which can be used to calculate directly the response of the structure by dynamic analyses. Note that the UHS curves for PSHA do not measure point-by-point the seismic waves and the material properties from a seismic source to a site.

6.2 Ground Displacement

Consider the ground displacement in space and time, u(x, t). The EG function for a differential operator $\mathcal{L}$ [u(x, t)] over the region Ω is defined to be a solution g(x, t; x', t') of $\mathcal{L}$ [g(x', t'; x, t)] $= \delta(x - x')\delta(t - t')$, by the Dirac function δ, that satisfies the boundary conditions $\mathcal{B}$ [u(x, t)]. The region Ω is the seismic sources, which can have a rectangular, square, circular, etc., area.

A particular solution of $\mathcal{L}$ [u(x, t)] $=$ f(x, t) can be obtained by performing a convolution integral:

$$\int_{0}^{\infty} dt' \int_{x' \in \Omega} f(x', t') g(x, t; x', t') dx' \tag{6.1}$$

where f(x', t') is the function that describes the seismic source at point x' and time t', and g(x, t; x', t') is the EG function. The convolution is a mathematical operation on two functions that produces a third function.

The ground displacement, u(x, t), under a seismic wave, at point x and time t (t > 0) can be defined as [55]:

$$u(x, t) = \int_{-\infty}^{\infty} \int_{0}^{\infty} f(x', t') g(x, t; x', t') dx' dt'. \tag{6.2}$$

---

[1] The use of the word "synthetic" instead of artificial or simulated, is purely to explain that the outputs in terms of displacements, velocities, accelerations, do not come from the recording of an actual earthquake. Thus, the use of one word rather than another does not have such a specific meaning.

Note that in Eq. (6.2), the source is limited in the axis x' and time t', therefore, the integral is calculated along a line instead of a region Ω as shown in Eq. (6.1).[2]

The use of the g(x, t; x', t') function provides the solution described as g:$\{\mathcal{L}[u(x, t)]$, $\mathcal{B} = 0$, u(x, t),(x, $-\infty$, ∞), t, (x', t')$\}$, where $\mathcal{L}[u(x, t)]$ represents the one-dimensional wave equation (described in Chap. 1) under unbounded media, $\mathcal{B} = 0$. Thus, the g solution is:

$$g = -\frac{1}{2}\Theta\big[\big(-t' + t\big) - \big(\big|-x' + x\big|\big)\big]$$

(6.3)

where $\Theta(x, t, x', t')$ is the Heaviside step function, and $|\cdot|$ represents the modulus.

The EG function only represents the medium where the effects of the propagation are registered. For the definition of a seismic source, events small enough are usually used in order that the frequency of interest is below the corner frequency of the source, $\bar{\omega}_c$, thus that the source of the EGF is a step function, which it is removed by deconvolution. Therefore, EGF is convolved with the source function.

In Appendix C in Chap. 8, the mathematical steps to obtain Eq. (6.3) are explained, which represents the solution of the wave equation through the Θ function. Thus, the problem should be to define the f(x', t') function, in fact, Eq. (6.2) contents the characteristics of the source correlated to the energy released by the earthquake, but it does not describe the ground motion. To consider the physical characteristics, the f(x', t') function can be correlated to the seismic moment, M_0, which expresses directly the earthquake energy (discussed in Chap. 1).

In this way, in the model there is consistency with a possible motion of the energy propagation, but the consistency with the rupture is not verified since the source could not simulate well the mechanism failure. One solution consists in defining a nonhomogeneous term for f(x', t') to describe the seismic source function in x' dimension, and time, t', distributed along the fault. However, this is not a simple task, for this in literature it is possible to find different solutions [57].

Probably, the most famous model to simulate the seismic source is the Brune model established firstly for a near-field, i.e., for displacements of the soil near to the seismic source as [51]:

$$v(t') = k\left(1 - e^{-\frac{t'}{\bar{t}}}\right)\Theta(t'), \quad with \; k = \frac{v_s \tau_0 \bar{t}}{\mu}$$

(6.4)

where k is a constant estimated in function of the propagation velocity of the s-wave, v_s, initial shear stresses τ_0 of the soil, μ is the Lamé constant at the source (equivalent to the shear modulus of the soil), and $\bar{t}$ can be considered the characteristic time of the failure (defined by $\bar{t} \approx a/v_s$, where a is the radius of the seismic source Ω if modelled as a circular area).

---

[2] Here the fault that has two or three-dimensions it is reduced into a line source. This would help to develop mathematically Eq. (6.2); however, it is possible that this approximation loses the physical nature of the source.

For a far-field, Eq. (6.4) can be also valid by considering some additional effects due to a large distance, as for instance the diffraction of the waves to be accounted for the phase, $t' - r/v_s$, where r is the site-source distance, and for the radiation damping of the waves to be accounted for the ratio, a/r. Thus Eq. (6.4) becomes:

$$v(r, t') = \frac{a}{r} \frac{v_s \tau_0}{\mu} \left(t' - \frac{r}{v_s} \right) e^{-\overline{\omega}_c \left(t' - \frac{r}{v_s} \right)} \Theta \left(t' - \frac{r}{v_s} \right) \tag{6.5}$$

where $\overline{\omega}_c$ is the corner circular frequency of the earthquake, which quantifies the seismic source and depends on radius, a, and velocity, v_s. This frequency is a specific point between two lines that describe the fault displacements plotted in the Fourier spectrum in a frequency domain, $\overline{\omega}$.

By the Fourier transform it is possible to define the absolute spectral displacements by:

$$|V(r, \overline{\omega})| = \frac{a}{r} \frac{v_s \tau_0}{\mu} \frac{1}{\overline{\omega}_c^2} \left[\frac{1}{1 + \left(\frac{\overline{\omega}}{\overline{\omega}_c} \right)^2} \right] = \overline{\Omega}_0 \left[\frac{1}{1 + \left(\frac{\overline{\omega}}{\overline{\omega}_c} \right)^2} \right], \tag{6.6}$$

which $\overline{\Omega}_0$ is the key parameter that accounts for the main information of the seismic source; thus, it could be useful correlate it to the mentioned seismic moment, M_0. In this sense, by some approximations the following Eq. (6.7) represents the Brune model providing the spectral displacements for a far-field:

$$|V(r, \overline{\omega})| = \frac{M_0}{4\pi r \rho v_s^3} \left[\frac{1}{1 + \left(\frac{\overline{\omega}}{\overline{\omega}_c} \right)^2} \right] \tag{6.7}$$

where ρ is the density of the soil.

Note that Eq. (6.7) provides a direct relation between M_0 and the displacement of the soil at a certain distance from the seismic source. This relation is important since it would allow to find unknown functions, $f(x', t')$, to solve the problem.

For instance, the analytical solution of Eq. (6.2), over N elements of area A ($\equiv \Omega$), where the amplitude of the displacement of the ground, u, is described by M_0, can be defined as:

$$u(x, t) = \sum_{i=1}^{N} \left(\frac{\mu A S(t')}{M_0} \right)_i g_n(x, t; x', t')_i \tag{6.8}$$

where $S(t')$ is the slip function at an element analytically deconvolved with the step function, and g_n is the EG function obtained from the recording of a small earthquake with a step source time function and interpolated to have a source and origin time at the location of the i-th element (for more details see [55]).

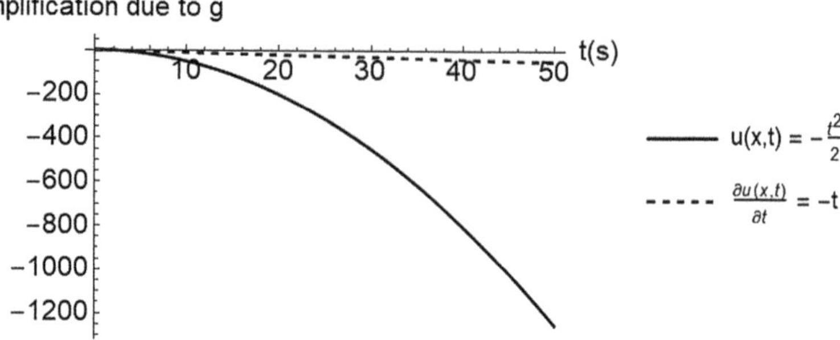

Fig. 6.2 Amplification of g function in t-domain

In Eq. (6.8) it is clearer that the model considers the medium as elastic-linear and isotropic thought μ constant.

By substituting Eq. (6.3) in Eq. (6.2) with $f(x', t') = 1$, it is possible to quantify the amplification of the Green's function, g, for the displacement and velocity, as shown in Fig. 6.2 (developed by Wolfram Mathematica software).

Figure 6.3 shows two examples of the ground displacements in function of the distance, x, and time, t. The outputs of this method provide directly the response of the soil, in terms of displacements, velocities, and accelerations, in the space and time. Note that the analysis is three-dimensional.

The examples are purely qualitative and the differences between both examples mainly regard the adopted $f(x', t')$ function; in Fig. 6.3a) the adopted function is more irregular than that in Fig. 6.3b).

It is possible to see that seismic waves alter the soil from a liner source distributed along $-\infty < x' < \infty$ (or $-\infty < x < \infty$) in accordance with Eq. (6.2). Note that Eq. (6.3) allows to convert a function with 4 variables to a function with 2 variables.

Despite the method can be considered advanced, the EG functions do not include the effects of the material nonlinearity, damping reductions, etc., therefore the synthesized ground motions are valid for linear responses. Like real and artificial accelerograms, these results can be modified and scaled to consider the mentioned effects.

Fig. 6.3 Example of results in terms of soil displacements (adapted from [57])

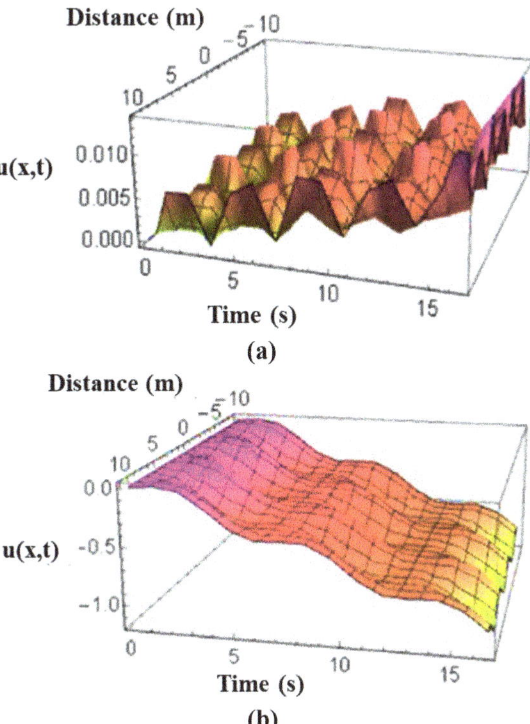

Chapter 7
Case Studies

Abstract In this chapter, some practical examples of technical and research projects are described. The studied structures include a pile-supported wharf, a fire station building, embankment and concrete dams, and storage tanks. The main mechanical concepts and analyses of the structures under earthquakes have been explained for didactic purposes; however, the goal is to show what seismic input was applied and to discuss some issues related to seismic design. For all examples, the authors participated in the design; in this sense, they represent professional experiences.

In this chapter some practical examples, regarding the behaviour and response of a structure under an earthquake, are described. For each example the main data, parameters, models, etc., are shown.

The main mechanical concepts and analyses, in a relative general way, have been explained for a didactic purpose. The idea is to show what method to estimate the seismic input has been used (the main topic of this book) for a specific structure and to discuss some consequences and issues regarding the seismic design. This book does not cover specific structural modelling and design; therefore, for these aspects, it is necessary to consult adequate books, codes, and manuals.

For all examples the authors participated to the analyses and design, in this sense they represent personal and professional experiences. Some case studies are technical projects whereas other ones are research projects.

7.1 Case Study 1: Pile-Supported Wharf

The studied pile-supported wharf is a new structure that forms part of the port placed at Northern of Venezuela, mainly intended for the manufacture and repair of oil tankers. Figure 7.1 shows a concept of the shipyard layout to frame the pile-supported wharf in a global way.

The first author of this book has worked in this technical project during the years 2013–2014, in the Brazilian company EGT Engineering in São Paulo.

E. Zacchei and R. M.L.R.F. Brasil, *Calculation of Seismic Actions for Structural Analysis*, https://doi.org/10.1007/978-3-032-08884-0_7

Fig. 7.1 Concept of the shipyard layout (preliminary version in 2010)

Figure 7.2 shows the technical drawing of the pile-supported wharf (Fig. 7.2a) using "AutoCAD" software, and its model by finite element method (FEM) using "Sap2000" software[1] (Fig. 7.2b). The drawing in Fig. 7.2a is only to show the general configuration without having to explain all the parameters and dimensions.

In Table 7.1 some important geometrical and mechanical parameters are listed to describe the structure.

The project appears very interesting and for some aspects it was complicated to find a better solution. In particular, two aspects are discussed here:

(i) The first one regards the necessity to develop the non-linear static pushover analysis since, due to the very high importance of the structure, this analysis is the better to estimate the non-liner response thus to quantify the real performance of the structure under an earthquake. This analysis can avoid the overdesigning of the structure.

(ii) The second one regards the difficult in estimating the real soil-structure interaction during an earthquake. It is very complicate to simulate the real soil behaviour under dynamic cyclic actions; also, complex liquefaction effects were possible. For this, two models have been developed to estimate a reliable range of the stiffnesses.

The pushover analysis is not considered a difficult analysis however some critical steps must be well defined. Substantially, horizontal external loads are applied to the structure and its horizontal displacements are registered. The results consist in plotting the elastic response spectrum by code (method used here and explained in Chap. 2) and the capacity curve (i.e., pushover curve), which defines the non-liner

---

[1] For an important project or, more generally, when there is financial and time availability to develop a project, it is good practice to model a structure using more than one software; this is to verify possible conceptual errors, to control better the outputs, to have more models available for different purposes, etc. Here, for brevity, only one software has been shown.

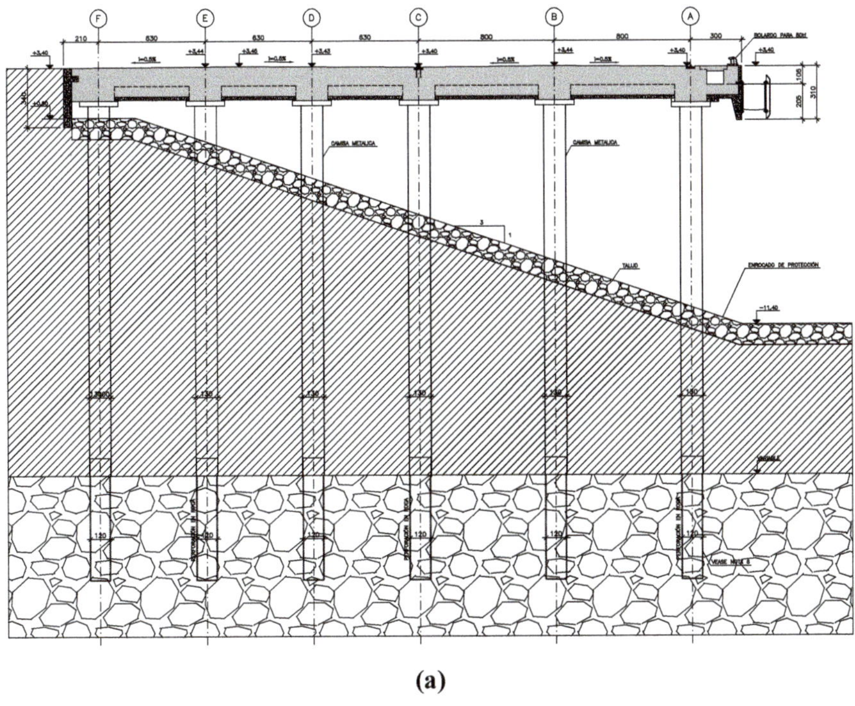

(a)

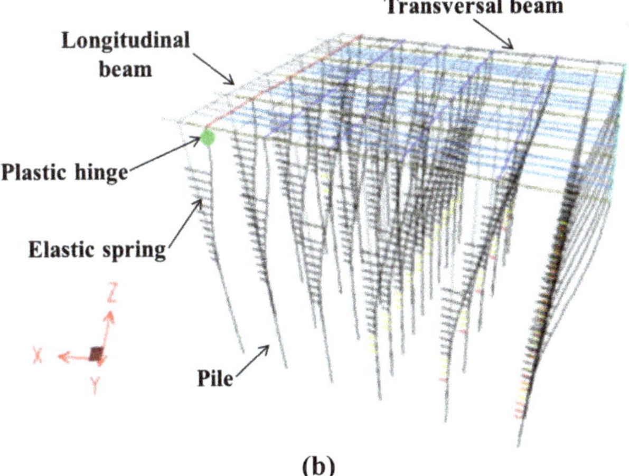

(b)

Fig. 7.2 Pile-supported wharf: **a** Technical drawing in transversal view in x-direction (some information is in Portuguese, however the most important ones for the example are explained in the text) and **b** FEM model (adapted from [27])

Table 7.1 Geometrical and mechanical parameters of the pile-supported wharf

Parameter	Value
Geometrical characteristics	
Wharf transversal length (x-direction)	40.0 m
Wharf longitudinal length (y-direction)[a]	63.0 m
Mean slab thickness (transversal slope of 0.50%)	0.40 m
Pile length	~ 25.0 m
Pile external diameter	1.20–1.30 m[b]
Mechanical characteristics	
Density concrete	25.0 kN/m^3
Compressive strength of the concrete	51.0 MPa
Elastic modulus of the concrete	29.52 GPa
Yield strength of the steel [c]	269.0 MPa
Elastic modulus of the steel [c]	206.0 GPa

[a] Basically, the height of the longitudinal and transversal beam is 1.90 m and 1.85 m, respectively
[b] This value can vary due to the presence of the metal jacket along the pile to avoid corrosion phenomena
[c] The parameters of the steel only refer to the metal jacket

response of the structure in terms of forces-displacements. The intersection point between both curves defines the performance of the structure.

To obtain this curve is necessary to establish a non-linear model for the material, the points (called plastic hinges) where this model must be applied, and the vertical and horizontal forces acting to the structure. Thus, the displacements are registered at a control point (coincident with the centre of the masses of the deck).

The formation of plastic hinges starts from the pile heads closest to the land (left side in Fig. 7.2a) and moves towards those closest to the sea (right side). In Fig. 7.2b the creation of the first plastic hinge by the FEM model is shown.

Here the classical Mander's model for confined concrete material (detailed in [18]) has been applied in several lumped plastic hinges placed at the points where the bending moment reaches the maximum value (i.e., at top of the piles). The considered horizontal seismic loads were the deadweight, accidental weights, pile weights, and hydrodynamic weights for a total of 13.28 MN. This total weight has been applied for each sub-modulus of longitudinal length of 7.0 m (for more details see [18, 27, 63, 64]).

Table 7.2 lists data regarding the soil and the seismic action. The geotechnical values correspond to the elastic springs (see Fig. 7.2b) applied at the piles to simulate the soil-structure interaction. The soil is basically formed by two different layers, i.e., clay and rock. The peak ground acceleration (PGA) indicates a very high seismic hazard at the site.

Figure 7.3 shows the results of the pushover analysis in terms of shear forces generated by the earthquake registered to the control point in function of the displacements.

Table 7.2 Geotechnical and seismic data

Parameter	Value
Geotechnical characteristics	
Stiffness of the clay layer	4900.0 kN/m^3
Stiffness of the rock layers	39,200.0–49,000.0 kN/m^3
Seismic characteristics	
PGA	0.43g
Spectral acceleration with 5.0% damping (for $T_1 = 1.64$ s)[a]	0.39g
Spectral acceleration with 5.0% damping (for $T_2 = 1.47$ s)	0.44g

[a] The parameters T_1 and T_2 are the first and second vibration mode, respectively, estimated by FEM analyses (g $= 9.81$ m/s^2)

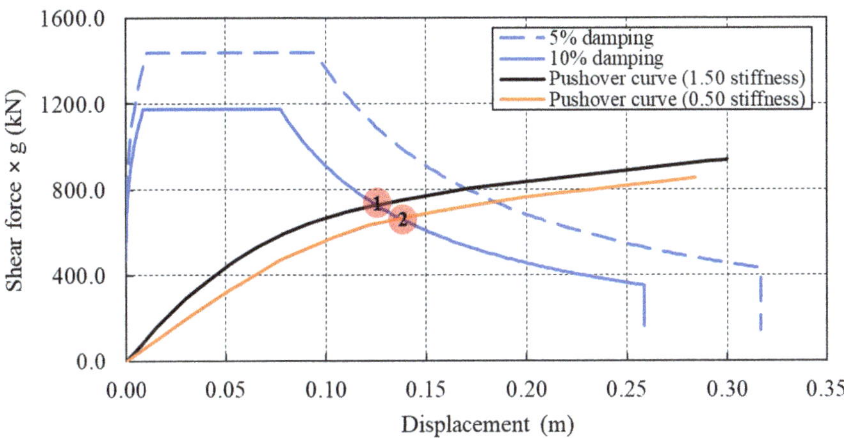

Fig. 7.3 Results of the pushover analyses highlighting two performance points (adapted from [18])

The displacements refer to the horizontal ones (x-axis in Fig. 7.2b) where the modal participating mass of the structure is greater (i.e., 99.76%),[2] thus in this direction the maximum response is obtained. This percentage refers to T_2 period value listed in Table 7.2.

The elastic spectrum has been reduced with a total damping ratio of 10.0% (including the hysteretic behaviour of the material) providing two performance points: (i) Point 1 for 1.50 stiffness, 12.54 cm and 7136.77 kN; (ii) point 2 for 0.50 stiffness, 13.74 cm and 6513.34 kN. Note that, for convention, the elastic spectrum is plotted by a force–displacement curve.

---

[2] The modal participating mass provides a measure of the real mass of the structure involved during the motion in a certain direction. Due to the supports applied at the structure to maintain the static equilibrium, some structural elements may not participate to the movement, thus their percentage (%) is usually less than 100.0%. This percentage is estimated by the modal analysis that also provides the vibration modes.

It is possible to see a slop change of the pushover curve at 6.67 cm (for 1.50 stiffness) and 8.66 cm (for 0.50 stiffness). This change indicates the creation of the first plastic hinge on the structure (see Fig. 7.2b), the others are formed subsequently; however up to the performance points the structure maintain its satisfactory resistance under the earthquake.

Thus, the global ductility demand in terms of displacements for both models can be measured and verified by $13.74/8.66 = 1.58 < 5.0$ (for 0.50 stiffness) and $12.54/6.67 = 1.88 < 5.0$ (for 1.50 stiffness) [78]. This ductility demand is a measure of the imposed post-elastic deformation on the structure.

These performance points represent the allowed maximum displacement by considering the elastic and plastic behaviour of the material. The main advantage of this analysis regards the fact that lower shear forces can be used for design, since by using a linear elastic curve the performance points would provide greater values.

As mentioned, due to the difficult to estimate the real soil-structure dynamic interaction, a range of stiffnesses has been considered, i.e., 0.50–1.50. Despite this range appears large, the difference between both curves does not appear very relevant guaranteeing the reliability of the results.

Due to the high structure flexibility and its configuration, torsional effects are possible thus both performance points have been amplified with a dynamic magnification factor of 1.30–1.45 accounting for the torsion, obtaining 17.86 cm ($= 1.30 \times 13.74$ cm) and 18.18 cm ($= 1.45 \times 12.54$ cm).

Finally, for a ductile structure subjected to high local ductility demand, the verification, that deformation demands are safely lower than the capacities of the plastic hinges, should be performed by comparing plastic hinge rotation demands, $\theta_{p,E}$, to the relevant design rotation capacities, $\theta_{p,d}$, i.e., $\theta_{p,d} > \theta_{p,E}$ [77].

The $\theta_{p,d}$ values have been calculated from ultimate curvatures by considering the safety factor of 1.40 that reflects local defects of the structure and the uncertainties of the model. For both models, the following values have been verified: 0.026 rad > 0.008 rad (for 0.50 stiffness), and 0.026 rad > 0.011 rad (for 1.50 stiffness).

7.2   Case Study 2: Fire Station Building

The fire station building is an existing reinforced concrete structure, built in 1984, placed at Penela city, Portugal. The aim of the technical project was to evaluate its stability and resistance for seismic retrofitting in the presence of a new metal construction built above, mainly intended for offices and dwellings. The first author of this book has worked in this project during the years 2019–2020, in the Portugues institute ITECONS in Coimbra.

The general idea was designing the building as a new structure and compare it with the information collected in situ (e.g., geometry of the concrete elements, quantity of steel bars, etc.). Note that this is an important task for this purpose and, at the same time, it is a relatively simple since destructive inspections were allowed only in some part of the structural elements, therefore superior longitudinal steel bars for beams and steel bars in the slab were unknown. Also, a destructive technique reduces the cross-section of the elements thus their resistance, therefore some caution is required when collecting information.

Figure 7.4 shows the existing structure (Fig. 7.4a) and its FEM model by using "SCIA Engineer" software indicating the mode for the fundamental period, T_1 (Fig. 7.4b). Piles 1–3 are shown to understand the spatial references of the structure. Table 7.3 lists the main geometrical and mechanical parameters.

In this case study, it could be interesting to highlight two aspects:

(a)

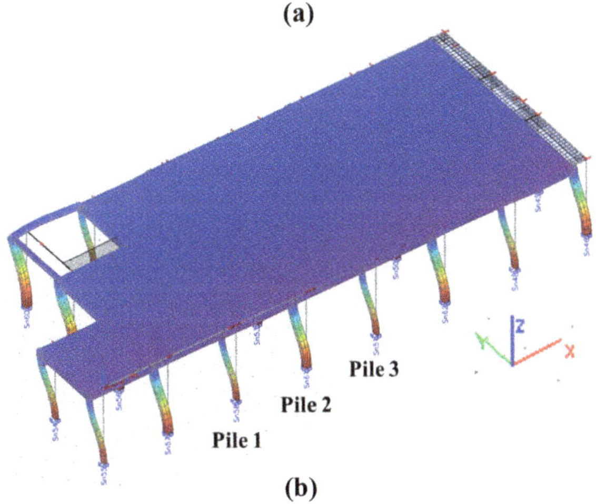

(b)

Fig. 7.4 Fire station building: **a** Real photo; **b** FEM model, indicating the first vibration mode (x-direction)

Table 7.3 Geometrical and mechanical parameters of the fire station building

Parameter	Value
Geometrical characteristics	
Slab area	381.17 m^2
Slab height	0.25 m
Beam lengths	4.50–9.50 m^a
Pile length (24 piles)	6.0 m^b
Mechanical characteristics	
Density concrete	25.0 kN/m^3
Compressive strength of the concrete	16.67 MPa
Elastic modulus of the concrete	29.0 GPa
Yield strength of the steel bars	348.0 MPa

[a] It varies for transversal or longitudinal beams, with cross-section areas between 0.05 and 0.37 m^2 (see also Table 7.5)
[b] This value considers the real pile length plus 1.0 m to consider the foundation, with cross-section areas between 0.07 and 0.17 m^2

(i) Despite the structure has a small dimension, it must be analysed with a particular attention since it is a public strategic structure with a very high importance class (i.e., buildings whose integrity during earthquakes is of vital importance for civil protection).

(ii) In Portugal from November 2019 it was mandatory to carry out the vulnerability analysis for existing structures in accordance with the methods (3 methods) described in [65], thus the traditional Portugues codes [28, 66] were insufficient during the project period; this was a coincidence that led to further analyses not initially planned.

The considered vertical loads are the deadweight of the structural elements, and non-structural elements (0.74 kN/m^2), and the accidental weights (2.0 kN/m^2). The weights due to the new metallic structure have been applied by concentrated loads in three directions (with values between 15.98 and 76.69 kN; these values include other actions as wind and snow).

The horizontal seismic action has been applied to the centre of the masses of the slab (due to a geometric asymmetry in the plane x, y, an eccentricity of 2.09 m has been considered). By considering the seismic data listed in Table 7.4, the seismic action was 847.0 kN (i.e., 345,400.0 kg × 0.25g, where 345,400.0 kg is the total mass of the building). In this case study the elastic response spectrum method has been used (Chap. 2).

The mentioned first aspect produces a seismic action amplified by an importance factor of 1.50, such that almost all the elements must be reinforced. This aspect caused a relevant increment of the costs making difficult the final realization. Actually, this factor is conservative and in favours of the safety overdesigning the whole structure.

Table 7.5 lists the comparisons between existing and necessary steel areas for each structural element. These differences are mainly due to the more simplified criteria

Table 7.4 Seismic data

Parameter	Value
PGA	0.07g
Importance factor	1.50[a]
Damping ratio	5.0%
Spectral acceleration (for $T_1 = 0.39$ s)[b]	0.25g

[a] Value referring to a structure with a very important class. Design PGA = $1.50 \times 0.07g = 0.11g$
[b] It refers to a modal participating mass of the structure of 95.10% in x-direction (Fig. 7.4b)

Table 7.5 Area of the steel bars (existing versus necessary)

Structural element		Longitudinal steel bars		Steel stirrups (2 legged)	
		Existing (cm^2)	Necessary (cm^2)	Existing (cm^2/m)	Necessary (cm^2/m)
Pile 40 × 43 cm^2	Top	12.56	18.57	2.26	4.30
	Bottom	12.56	20.64	2.26	4.30
Pile 20 × 35 cm^2	Top	3.14	15.45	1.41	3.50
	Bottom	3.14	15.12	1.41	3.50
Pile 40 × 35 cm^2	Top	9.04	18.81	2.26	4.0
	Bottom	9.04	18.48	2.26	4.0
Beam 26 × 20 cm^2	Superior (support)	N/A	2.51	N/A	2.60
	Inferior (middle)	3.39	1.56	1.41	2.60
Beam 35 × 20 cm^2	Superior (support)	N/A	5.59	N/A	6.0
	Inferior (middle)	3.39	2.49	1.41	3.50
Beam 43 × 85 cm^2	Superior (support)	N/A	33.35	10.05	11.25
	Inferior (middle)	12.56	9.98	3.35	4.30

Structural element		Reinforcement steel mesh	
		Existing (cm^2/m)	Necessary (cm^2/m)
Slab	Superior	N/A	3.0
	Inferior	6.64	12.0

Note N/A = Not available value

present in the old code (adopted to originally design this structure) with respect the modern one, and to the neglecting (more probably) of the seismic actions.

Regarding the second aspect, in accordance with [65], a fire station building should be verified by using the "method III" (based on the "yield criteria", where a pushover analysis is advisable), however given that the following requirements were respected, a method II was sufficient:

(i) The number of floors is less than 4.0, and the area of the building is less than 400.0 m^2; (ii) the structure is considered regular in height and in plan. The building does not have slenderness in plan; (iii) the building does not interact with adjacent buildings; (iv) the stratigraphic profile was assumed as a soil "type C" (i.e., dense or medium-dense sand soil [14]).

This contradiction in using method II or III is correlated to the fact that the structure is important but small. Results, by using method II (substantially based on the static equilibrium of the force, analogue to the simplified method in Sect. 3.1), shown that the building was verified in a global way, however not guaranteeing that each element is also verified. This would indicate that that most of the building elements need to be reinforced as shown in Table 7.5. The conclusion is that for small but strategic structures, the use of importance factors by coefficients and of simplified analysis methods can lead to overdesigning of the structure.

7.3   Case Study 3: Embankment Dams

This case study regards the estimation of the vertical settlements of the embankment dams under earthquakes. The first author of this book has worked in this technical-scientific project during the years 2023–2024 in the national working group of the ITCOLD, Italy [67].

This estimation is based on the simple analytical model described in [68, 82] (Fig. 7.5a), which substantially consists in defining the normalized crest settlement (NCS), in function of geometrical parameters by Eq. 7.1, or in function of PGA acceleration and moment magnitude, M_w, of the considered seismic event by Eq. 7.2, both expressed in percentage:

$$\text{NCS} = \left(\frac{\text{CS}}{\text{DH} + \text{AT}} \right) 100 \tag{7.1}$$

$$\text{NCS} = e^{[(5.70\,\text{PGA})+(0.471\,M_w)-7.22]} \tag{7.2}$$

where CS is the vertical settlement of the dam crest, DH is the dam height, and AT is the alluvium thickness where the dam is placed.

Equation (7.1) describes the simplified model, which can be used a-posteriori (i.e., after an earthquake) when the CS value is known, whereas Eq. (7.2) appears more useful estimating the NCS a-priori thus providing preliminary results. The constants in Eq. (7.2) have been defined by a mathematical regression of the collected data regarding several real embankment dams. The choice of one relation over the other depends exclusively on the available data.

Thus, the crest settlement would represent the unique index for estimating the global behaviour of the dam and its performance under an earthquake. By this index

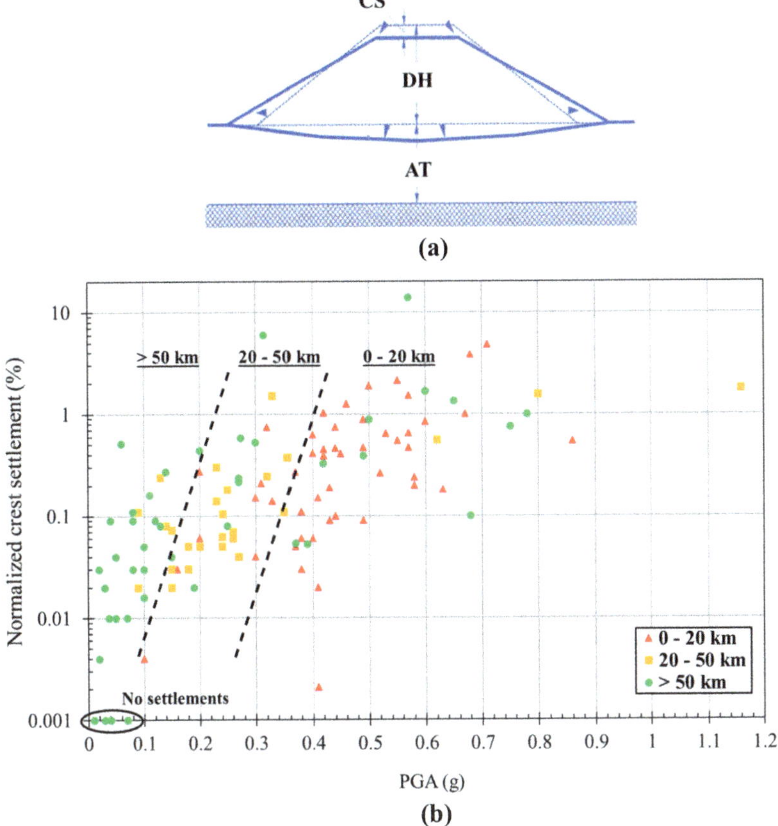

Fig. 7.5 Crest settlement of embankment dams: **a** Model; **b** results (adapted from [68])

it is possible to classify the damage level of the dam in serviceability limit states (SLSs), i.e., operational and damage, and ultimate limit states (ULSs), i.e., life safety and collapse.

This technical-scientific project consists in increasing and improving the existing database by considering the relevant cases in the World. For this, several technical reports, academic papers, bulletins, etc. have been retrieved, analysed and discussed during the meetings among the specialist members of the working group. This could represent the interesting aspect to be highlighted in this case study.

Figure 7.5 shows the mentioned simplified model (Fig. 7.5a) and the results (plotted in a logarithmic scale) of all collected dams in terms of NCS values, PGAs and dam-epicentre distances (Fig. 7.5b).

It is evident that this model appears very practical; however, it neglects possible settlements in other directions, more complex failure mechanisms, etc. This model should be used only for a preliminary estimation.

Here, the used method to estimate the seismic input is the time-history representation (Sect. 3.2.1), since the PGA values have been retrieved directly from real accelerations.

For each dam, in addition to the mentioned parameters, other ones have been collected, as for instance, the dam's name, the location where the dam placed, its construction year, the dam type, the length of the dam crest, etc., and the information regarding the earthquake.

Thus, by considering the existing database (82 case studies listed in [82]) plus 41 new case studies analysed in this project (new embankment dams), more than 1500.0 parameters have been controlled and analysed. In Fig. 7.5b, it is possible to see all 123.0 embankment dams.

It was noted that a better correlation for NCS-PGA points is plotted by also considering the dam-epicentre distance, since as expected the vertical displacement, thus the damage, decreases by increasing this distance. For this, three limits of distances have been defined to quantify their influences (i.e., 0-20 km, 20-50 km, > 50 km).

Results shown that 47.0% of the dams reported no detectable damage (under a mean value of PGA $\approx$ 0.20g), 21.0% exceed the operational limit state (PGA $\approx$ 0.34g), 17.0% exceed the damage limit state (PGA $\approx$ 0.44g), and 13.80% exceed the ultimate limit state (PGA $\approx$ 0.61g). In general, results shown that embankment dams could withstand seismic accelerations up to 0.20g without significant damage, thus a value of PGA = 0.20g can be considered as a lower bound for SLS, whereas PGA = 0.30g can be considered as the lower bound for ULS in favour of safety.

Finally, PGAs greater than 0.70g have been recorded only for earthquake magnitudes greater than 6.7 producing vertical settlements concerning SLS (damage) and ULS. The collapse of the dams (1.20%) has been reached only for earthquake magnitudes greater than 7.6.

7.4 Case Study 4: Concrete Dam

The Rules Dam is a double-arch concrete dam placed at Granada, Spain, built in 2003, mainly intended for supply, irrigation and flood control. The authors of this book have worked in this research project during the year 2016–2018, in the Polytechnic School of the University of Salamanca (USAL), Avila, Spain.

Figure 7.6 shows a real photo (Fig. 7.6a) of the Rules Dam and its model (Fig. 7.6b) by finite element method (FEM) using "Sap2000" software.

Table 7.6 lists the geometrical and mechanical characteristics of the dam (for more details see [69]).

Along the crown length, vertical joints each 20.0 m have been placed providing 32.0 blocks, as shown in Fig. 7.6b. Also, longitudinal joints have been placed, between the foundation and the height of 50.0 m below the crest, to reduce the

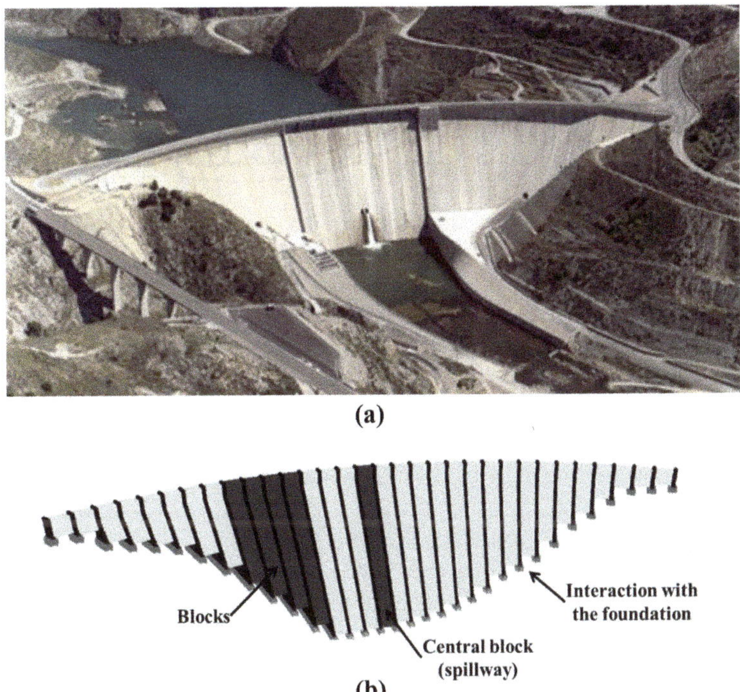

Fig. 7.6 Rules Dam: **a** Real photo (retrieved from [69]); **b** FEM model

Table 7.6 Geometrical and mechanical parameters of the concrete dam

Parameter	Value
Geometrical characteristics	
Height	130.0 m
Crown length	620.0 m
Crown width	10.0 m
Concrete volume	2032.0×10^3 m^3
Radius[a]	500.0 m
Mechanical characteristics	
Density concrete	24.0 kN/m^3
Characteristic compressive strength of the concrete (at 90 days)	17.16 MPa
Fundamental structural period	0.284 s[b]

[a] It refers to the longitudinal view along the crown length. In a cross-section view, the downstream and upstream slope faces are 0.60/1 and 0.18/1, respectively

[b] Value estimated by authors using the FEM model [45, 60, 61, 70]

size of the larger blocks to facilitate the heat dissipation during the concrete hardening process (the purpose was to maintain a longitudinal length of the central blocks less than 50.0 m) [79].

For the monitoring and maintenance of the operation and construction (including consolidation grouting, drains, etc.) of the dam, an extensive network of galleries has been built (15 longitudinal galleries at different heights in the dam body).

By geophysics in-situ surveys, the Rules Dam must support on a rock with an average shear wave velocity greater than 2000.0 m/s (very rigid rock), where, as discussed in Chap. 2, the amplification phenomena due to site effects do not occur.

In general, for concrete dams the stability problem under an earthquake could not be an interesting issue since these types of dams have a very relent mass that guarantees their structural stability. Due to the relevant mass and its deformation mode (the crest deforms towards downstream), a concrete dam could be simulated as a simple oscillator (introduced in Chap. 2), however this idealization could be very approximate and questionable. It could be of interest to study the creation of cracks in the dam body, the damage of some sensible parts (e.g., spillway, crest, etc.), their global functionality during an earthquake, etc. [62].

The Rules Dam is placed in an area with a very high seismic hazard. Given that Rules dam was built in 2003, it could be interesting to define some controlling earthquakes accounting for the seismic hazard well defined by the Spanish ZESIS [42] database published in 2015, thus 12 years after the dam construction. As discussed in Sect. 5.2, these controlling earthquakes serve to define the safety evaluation earthquake (SEE), which represents the maximum level of ground motion for which the dam should be designed or re-verified (if necessary). The controlling earthquakes are defined for the deterministic seismic hazard analyses method.

This dam is situated among three seismogenic zones (ZS), ZS 35, ZS 36, and ZS 38. From these ZSs it is possible to select the following controlling earthquakes:

(i)   "Arenas del Rey" earthquake of year 1884, with a moment magnitude M_w 6.5, probably generated by a normal fault as predominant focal mechanism (see Chap. 1) of the ZS 35.
(ii) "Baza" earthquake of year 1531, with a moment magnitude M_w 6.2, probably generated by a normal fault as predominant focal mechanism of the ZS 36.
(iii) "Alhama de Almeria" earthquake of year 1522, with a moment magnitude M_w 6.5, probably generated by tears with variable oblique component as predominant focal mechanism of the ZS 38.

All three events are the largest in terms of magnitude registered in the mentioned ZSs, also they are superficial thus they can produce the maximum level of ground motions. They have been considered to define the respective seismic hazard where a specific focal mechanism is defined.

Obviously, these events are not the unique possible controlling earthquakes, since they only consider the highest magnitude partially neglecting the dam-epicentral distance, which is an important factor to be considered. With a distance value it is possible to estimate the acceleration to be applied to the dam body by an adequate attenuation equation.

7.5 Case Study 5: Storage Tanks

This case study regards storage tanks under earthquakes. The authors of this book have worked in this research project during the years 2015–2016, in the Polytechnic School of the University of São Paulo (USP), Brazil.

Storage tanks are usually placed in industrial plants to storage water, petroleum, gas, etc. The first author was coordinator to design new parts of the industrial plant for storage oxygen shown in Fig. 7.7, during the years 2014–2015 in the Brazilian company CFPS Engineering in São Paulo. This is a purely an example only to frame a storage tank since this plant was not subjected to earthquake actions.

The theoretical model to evaluate the dynamic behaviour of an anchored tank on-ground is described in literature [71]. The motion of the fluid contained in a rigid cylinder can be expressed substantially as the sum of two separate contributions, i.e., rigid impulsive, and convective. The rigid impulsive component (expressed by impulsive pressures) satisfies exactly the boundary conditions at the walls and the bottom of the tank. It gives zero pressure at the original position of the free surface of the fluid in the static situation, however in the dynamic response, these pressures are not null due to the presence of the waves described by the convective pressures also called "sloshing" phenomenon.

In this sense, the interesting aspect could consist in using the simple oscillator (adopted for structures as discussed in Chap. 2), also for a hydraulic problem. Thus, the tank-liquid system is modelled by two single-degree-of-freedom systems, one corresponding to the impulsive component moving together with the wall, and the other one corresponding to the convective component.

Fig. 7.7 Industrial plant for storage oxygens in Brazil (by Google maps)

Note that this model is valid for rigid and flexible tank. In the latter case, another pressure must be considered (flexible pressure) which satisfies the condition that the radial velocity of the fluid along the wall equals the deformation velocity of the tank wall, as well as the conditions of zero vertical velocity at the tank bottom and zero pressure at the free surface of the fluid [81].

The dynamic coupling between the sloshing and the flexible components is very weak, due to the large differences between the frequencies of the sloshing motion and of the deformation of the wall, which allows determining the third flexible component independently of the others. The rigid impulsive and the sloshing components remain therefore unaffected.

Figure 7.8a shows the theoretical model where, m_0 is the impulsive mass attached rigidly to the tank at the proper height, h_0, and m_1 is the convective mass regarding its fundamental mode of oscillation, which oscillates horizontally in x-direction against a restraining elastic spring, k_1, at height h_1. The oscillating water surface is described by the ws parameter, and d is the sloshing height measured from the liquid surface at rest. Figure 7.8b shows its FEM model using "Sap2000" software.

Here a parametric analysis has been carried out, with a h/R ratio that varies from a squat tank to a slender tank, where h is the height to the free surface of the liquid, and R is tank radius.

It is normally unconservative to consider the tank as rigid (e.g., concrete tank), thus in this example the walls of the tank are considered made by steel (density of $78.50\,\text{kN/m}^3$, modulus of elasticity of $210.0\,\text{GPa}$, and an equivalent uniform thickness of 8.0 mm). The liquid is petroleum (density of 9.0 kN/m^3), and the used PGA is

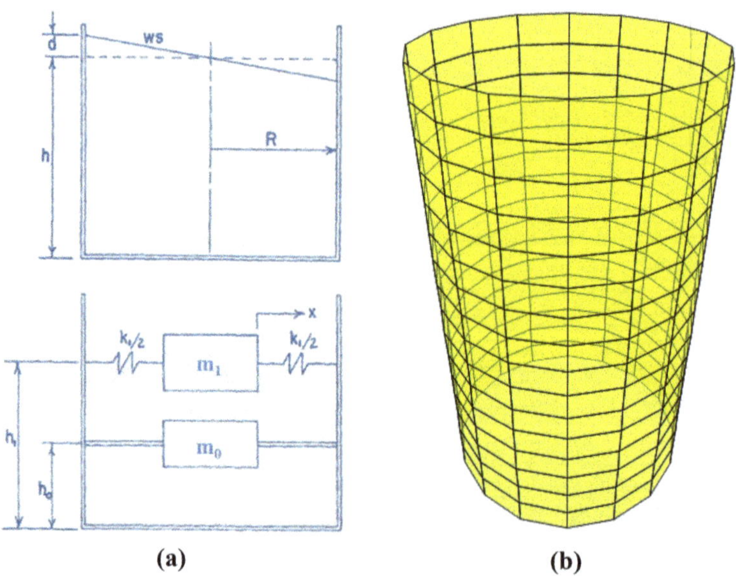

Fig. 7.8 Storage tank: **a** theoretical (adapted from [71]) and **b** FEM model

0.37 g [73, 74]. These values would regard, for instance, a flexible storage tank for an industrial plant placed in an area with a high seismic hazard.

The used method to estimate the PGA value is the elastic response spectrum reduced by a damping ratio of $\xi = 2.0\%$ for steel material and $\xi = 0.50\%$ for liquid (Chap. 2). The considered values vary between 0.20–1.15 g.

Table 7.7 lists the results of the seismic response for fixed base cylindrical tanks by using the simplified procedure shown in Eurocode [72], which consists in defining external shear forces, Q, and overturning moments immediately above, M, and below the base plate, M'. The former moments include only the contributions of the pressures on the walls, whereas the latter ones include the pressures on the walls and on the tank bottom.

The tank bottom pressures regard the whole system (i.e., tank and foundation). This dynamic fluid pressure is transferred directly either to the subgrade or to other supporting structural elements by increasing the effective vertical moment arm to the applied forces. For this reason, M' values are greater than M values.

The structural natural periods, T_0 and T_1, refer to the impulsive and convective mass, respectively. Note that in code [72], these values are estimated by specific coefficients that allow to adopt the simplified procedure, however they are analogue to that for the simple oscillator, for instance, for T_1 period the relation is $T_1 = 2\pi\sqrt{m_1/k_1}$. The d_{max} parameter is the peak height of the sloshing wave provided mainly by the first mode.

Table 7.7 Dynamic response of the tank-liquid systems by the simplified procedure

h/R	T_0 (s)	T_1 (s)	Q (MN)	M (MN m)[a]	M' (MN m)[a]	d_{max} (m)[b]
0.30	0.02	4.67	0.87	4.89	6.34	0.84
0.50	0.03	3.89	1.38	5.63	7.92	1.01
0.70	0.04	3.58	2.01	6.99	10.02	1.09
1.00	0.05	3.40	3.08	10.47	14.11	1.09
1.50	0.08	3.31	5.03	20.42	24.14	1.13
2.00	0.10	3.31	7.02	35.22	38.45	1.13
2.50	0.14	3.31	9.01	54.66	57.50	1.13
3.00	0.18	3.31	11.0	78.56	81.43	1.13

[a] Here also the mass of tank roof has been considered
[b] These values, plus height h, are lower than the assumed total height of the tank, i.e., 17.0 m, thus leakage of the liquid is avoided

Chapter 8
Appendices

Abstract For completeness, three appendices are provided to support the methodologies described in this book and to explain some more complex mathematical theories. In particular, they cover equations related to the Fourier spectrum, the elastic response spectrum with four branches, and the empirical Green's function.

8.1 Appendix A: Fourier Spectrum

To transform a function in a time domain to a frequency domain, and vice versa, it is possible to use the Fourier spectrum. It is given for the ground motion by [15]:

$$\ddot{V}_g(i\overline{\omega}) \equiv \int_{-\infty}^{\infty} \ddot{v}_g(t) e^{[-i\overline{\omega}t]} dt \tag{8.1}$$

where $\ddot{v}_g(t)$ is the ground acceleration in time, t, $\overline{\omega}$ is the circular frequency of harmonic forcing function, and i is the imaginary constant. This allows to express the ground acceleration through the superposition of a full spectrum of harmonics as indicated by the following inverse relation:

$$\ddot{v}_g(t) = \frac{1}{2\pi} \int_{-\infty}^{\infty} \ddot{V}_g(i\overline{\omega}) e^{[i\overline{\omega}t]} d\overline{\omega}. \tag{8.2}$$

Assuming that the ground motion is nonzero only in the range $0 < t < t_1$, Eq. (8.1) can be separated into its real and imaginary parts as follows:

$$\ddot{V}_g(i\overline{\omega}) = \int_0^{t_1} \ddot{v}_g(t) \cos(\overline{\omega}t) dt - i \int_0^{t_1} \ddot{v}_g(t) \sin(\overline{\omega}t) dt \tag{8.3}$$

with the Fourier amplitude spectrum defined as $\left|\ddot{V}_g(i\overline{\omega})\right|$ (expressed as a velocity), when divided by 2π, provides the ground acceleration amplitude per unit of $\overline{\omega}$. This quantity plus the Fourier phase spectrum, $\theta(\overline{\omega})$, can interpret various phenomena associated with the transmission of energy from the earthquake source to distant locations.

Note that this transformation regards only the earthquake and there is no information concerning the structure.

The advantages of the Fourier analysis are basically two:

(i) It allows to approximate a general dynamic load, as for example an irregular earthquake acceleration, by the sum of a series (called "Fourier series") of simple harmonic components. In this way, it is possible to obtain a mathematical expression to be put at right-hand side of the Eq. (2.3) in Chap. 2. Thus, this series transforms a real trend in an ideal approximated trend.

(ii) It transforms a spectrum in the time domain to a one in the frequency domain by the fast Fourier transform (FFT). An earthquake acceleration with an irregular trend has not a defined period (or frequency) thus it is supposed that the period tends to infinity in order to use the FFT.

8.2 Appendix B: Elastic Response Spectrum with Four Branches

The elastic response spectrum with four branches, correlated to the PSD function described in Chap. 4, is defined by:

$$S_a(T) = S_0\left[1 + (\alpha - 1)\frac{T}{T_B}\right], \quad 0 \le T \le T_B \tag{8.4}$$

$$S_a(T) = \alpha S_0, \quad T_B \le T \le T_C \tag{8.5}$$

$$S_a(T) = \alpha S_0\left(\frac{T_C}{T}\right)^{k_1}, \quad T_C \le T \le T_D \tag{8.6}$$

$$S_a(T) = \alpha S_0\left(\frac{T_C}{T_D}\right)^{k_1}\left(\frac{T_D}{T}\right)^{k_2}, \quad T \ge T_D \tag{8.7}$$

where S_0 is the peak ground acceleration, α is the dynamic amplification factor, and T is the structural period where assumes the value of T_B, T_C, and T_D as shown in Fig. 8.1. Note that this elastic response spectrum is the same explained in Chap. 2; some different symbols and acronyms have been used since Appendix B is consistent to the method explained in Chap. 4.

The exponents k_1 and k_2 provide the spectrum shape, thus they are correlated to the following exponents:

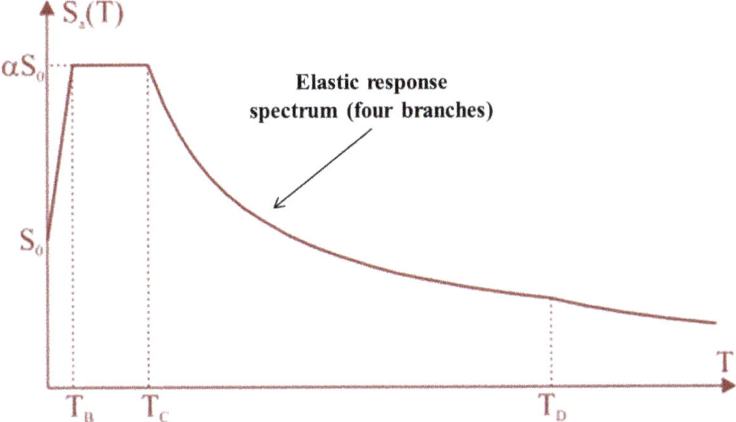

Fig. 8.1 Elastic response spectrum (adapted from [34])

$$e_1 = 2k_2 - 1 - L(\omega_D) \tag{8.8}$$

$$e_2 = 2k_1 - 1 - L(\omega_C) \tag{8.9}$$

$$e_3 = -1 - \gamma - \beta_2 L(\omega_C) \tag{8.10}$$

$$e_4 = -1 - \gamma - \beta_1 \left[L(\omega_B) + 2\left(\frac{\alpha - 1}{\alpha}\right) \right] \tag{8.11}$$

where the damping ratio of the structure, ξ, is included in γ parameter, and:

$$L(\omega_j) = 2\omega_j \frac{d(\log(\eta_U))}{d\omega_j} \tag{8.12}$$

$$\beta_1 = \left(\frac{\omega_C}{\omega_B}\right)^{e_3+1} \beta_2 + \left[1 - \left(\frac{\omega_C}{\omega_B}\right)^{e_3+1} \right] \frac{\gamma + e_3 + 1}{e_3 + 1} \tag{8.13}$$

$$\beta_2 = \left(\frac{\omega_D}{\omega_C}\right)^{e_2+1} \frac{\gamma + e_1 + 1}{e_1 + 1} + \left[1 - \left(\frac{\omega_D}{\omega_C}\right)^{e_2+1} \right] \frac{\gamma + e_2 + 1}{e_2 + 1} \tag{8.14}$$

where the circular frequency, ω_j, with $j = B, C, D$, is $\omega_j = 2\pi/T_j$, and L is a generic function (for more details see [15]). Other parameters have been already explained.

Equations (8.8)–(8.14) provide the analytical closed forms to find the PSD function compatible with an elastic response spectrum with four branches.

8.3   Appendix C: Green's Function

Consider an inhomogeneous wave equation (Eq. (1.2) described in Chap. 1) to be equal to a given function, $h(x, t) \neq 0$, in one dimension, x $(-\infty < x < \infty)$, and tine, $t > 0$. The empirical Green's function (EGF) $g(x, t; x', t')$, by the Dirac delta function δ, associated with this wave equation satisfies [57]:

$$\frac{\partial^2 g(x,\ t;\ x',\ t')}{\partial x^2} - \frac{1}{v_s^2} \frac{\partial^2 g(x, t; x', t')}{\partial t^2} = \delta(x - x')\delta(t - t') \qquad (8.15)$$

where v_s is the propagation velocity of the shear wave, s-wave. The point x' and time t' refer to the seismic source function, $f(x', t')$. The function g is a fourth-order elasticity tensor.

The EGF represents the response of a unidirectional unit-impulse under specific conditions. Assuming another Green's function, G, $G(x, t; x', t') = (e^{ixx'})r(x, t)$, whit a general function $r(x, t)$, Eq. (8.15) can be written:

$$\frac{1}{v_s^2} \frac{\partial^2 r(x, t)}{\partial t^2} + x^2 r(x, t) = \delta(t - t') \qquad (8.16)$$

therefore

$$\frac{1}{v_s^2} \frac{\partial^2 r(x, t)}{\partial (t - t')^2} = -x^2 r(x, t), \qquad (8.17)$$

which has solutions $r(x, t) = A(t) \sin [x(t - t')]$, with $A(t) = \int [\delta(t - t')/x]dt$, that is an amplitude of sine function.

Considering $\int \delta(t - t')dt = \Theta(t - t')$, $r(x, t)$ is:

$$r(x, t) = \frac{\sin[x(t - t')]\Theta(t - t')}{x} \qquad (8.18)$$

where the Heaviside step function, Θ, simulate the impulse (the function Θ is 1 only if none of the x, t, x', t' are not positive). Therefore, G, can be written as:

$$G(x, t; x', t') = \frac{e^{ixx'} \sin[x(t - t')]\Theta(t - t')}{x}. \qquad (8.19)$$

By using the inverse of Fourier's transform, F^{-1}:

$$g(x, t; x', t') = F^{-1}[G(x, t; x', t')] = \frac{1}{2\pi} \int_{-\infty}^{\infty} G(x, t; x', t')e^{-ix^2} dx$$

$$= \int_{-\infty}^{\infty} \frac{\Theta(t - t') \sin[x(t - t')]e^{ix(x'-x)}}{2\pi x} dx \tag{8.20}$$

and by using two general identities: (i) $2\sin(\alpha)[\cos(\beta) + \sin(\beta)] = \sin(\alpha - \beta) + \sin(\alpha + \beta) + \cos(\alpha - \beta) - \cos(\alpha + \beta)$, and (ii) $e^{i\alpha} = \cos(\alpha) + i\,\mathrm{sen}(\alpha)$, it is obtained:

$$g(x, t; x', t') = \frac{\Theta(t - t')}{4\pi} \int_{-\infty}^{\infty} \left(\frac{\sin[x(t - t' + x' - x)]}{x} + \frac{\sin[x(t - t' - x' + x)]}{x} \right.$$
$$\left. +i\frac{\cos[x(t - t' + x' - x)]}{x} + i\frac{\cos[x(t - t' - x' + x)]}{x} \right) dx. \tag{8.21}$$

Neglecting the imaginary part, knowing that $\int_{-\infty}^{\infty} \frac{\sin(\alpha)}{\alpha} d\alpha = \pi$, and introducing the sign function as $\mathrm{sgn}(\alpha) = \{-1, 0, 1\}$ for $\alpha < 0$, $\alpha = 0$, $\alpha > 0$, respectively, Eq. (8.21) is:

$$g(x, t; x', t') = \frac{\Theta(t - t')}{4\pi} \pi[\mathrm{sgn}(t - t' + x' - x) + \mathrm{sgn}(t - t' - x' + x)] \tag{8.22}$$

that, considering $\mathrm{sgn}(\alpha) \approx 2\,\Theta(\alpha)$, becomes Eq. (6.3) of Chap. 6.

References

1. M. Levy, M. Salvatori, *Perché gli Edifici Cadono* (Bombiani, Milan, Italy, 1997), p. 359
2. P. Odifreddi, *Come Stanno le Cose – Il mio Lucrezio, la mia Venere* (prose translation of De Rerum Natura by Titus Lucretius Carus) (Rizzoli, Italy, 2013) p. 311
3. S.L. Kramer, *Geotechnical Earthquake Engineering* (Prentice-Hall, Upper Saddle River, New Jersey, 1996), p. 653
4. E. Zacchei, R. Brasil, K-means for earthquakes: disaggregation analyses of small events by considering wave components and soil types. Arab. J. Geosci. **17**, 1–14 (2024)
5. G. Lanzo, F. Silvestri, *Risposta Sismica Locale – Teorie ed Esperienze* (Ed. Hevelius, Italy, 1999), p. 160
6. Geographic National Institute (IGN), Spain. Accessed on Nov 2024. https://www.ign.es/web/ign/portal/inicio
7. J. B. M. Bravo, *Teoría sobre la propagación de ondas sísmicas. Ondas Lg* (National Centre for Geographic Information - CNIG, 2017), p. 188
8. M. Assumpção, A regional magnitude scale for Brazil. Bull. Seismol. Soc. Am. **73**, 237–246 (1983)
9. United States Geological Survey (USGS) - Science for a changing world, seismic hazard maps and site-specific data (2022). Accessed on Nov 2024. https://earthquake.usgs.gov/earthquakes/eventpage/usp0009d4z/executive
10. R. Brasil, A.M. Da Silva, *Introdução à Dinâmica das Estruturas para a Engenharia Civil* (Edgard Blucher Ltda, São Paulo, Brasil, 2013), p. 268
11. L. Luzi, R. Puglia, E. Russo, ORFEUS WG5, Engineering Strong Motion (ESM) database, version 1.0 (National Institute of Geophysics and Volcanology - INGV, Observatories & Research Facilities for European Seismology, 2016). https://esm.mi.ingv.it
12. S. Timoshenko, J.N. Goodier, *Theory of Elasticity*, 2nd edn. (McGraw-Hill Book Company, Inc., 1951), p. 506
13. A. Chiaradonna, Defining the boundary conditions for seismic response analysis—a practical review of some widely-used codes. Geosciences **12**, 1–15 (2022)
14. European Committee for Standardizations (CEM), Eurocode 8: Design of structures for earthquake resistance—Part 1: General rules, seismic actions and rules for buildings, EN 1998-1:2004, Brussels, Belgium
15. R.W. Clough, J. Penzien, *Dynamics of Structures* (McGraw-Hill, New York, USA, 2003), p. 752
16. E. Zacchei, P. Lyra, Recalibration of low seismic excitations in Brazil through probabilistic and deterministic analyses: application for shear buildings structures. Struct. Concr. **1–19**, 2022 (2022)
17. Minister of Infrastructure and Transport, Norme Tecniche per le Costruzioni, NTC 2008, Rome, Italy, p. 372 (2008)

18. E. Zacchei, P.H.C. Lyra, F.R. Stucchi, Pushover analysis for flexible and semi-flexible pile-supported wharf structures accounting the dynamic magnification factors due to torsional effects. Struct. Concr. **1–20**, 2020 (2020)

19. J.M. Gere, Resistencia de los Materiales, 5th edn. (Editor Paraninfo S.A., 2009), p. 944

20. E. Zacchei, J.L. Molina, Introducing importance factors (IFs) to estimate a dam's risk of collapse produced by seismic processes. Int. J. Disaster Risk Reduct. **60**, 1–13 (2021)

21. K. Poljansek, M. Marin Ferrer, T. De Groeve, I. Clark, *Science for Disaster Risk Management 2017: Knowing Better and Losing Less* (Disaster Risk Management Knowledge Centre - DRMKC, European Union, Luxembourg, 2017), p. 554

22. Fundación Venezolana de Investigaciones Sismológicas Covenin 1756-1. Funvisis. Edificaciones Sismorresistentes, Parte 1: Requisitos. Fondonorma, Caracas (2001)

23. Servicio Nacional de Capacitacion para la Industria de la Construccion (SENCICO), Norma 0.30 Diseño Sismorresistente, Lima, Peru, p. 81, 2020.

24. IS 1893-1: Criteria for Earthquake Resistant Design of Structures, Part 1: General Provisions and Buildings, CED 39: Earthquake Engineering, India, p. 45 (2002)

25. Associazione Italiana Calcestruzzo Armato Precompresso (AICAP), Progettazione Sismica di Edifici in Calcestruzzo Armato, Guida all'Uso del Eurocidice 2 con riferimento alle Norme Tecniche D.M. 14.1.2008, Volume 1 e 2 (2008)

26. S. Timoshenko, J.M. Gere, *Theory of Elastic Stability*, 2nd edn. (McGraw-Hill International Book Company, Inc., 1985), p. 541

27. E. Zacchei, P.H.C. Lyra, F.R. Stucchi, Nonlinear static analysis of a pile-supported wharf. Ibracon Struct. Mater. J. **12**, 998–1009 (2019)

28. Regulamento de Solicitações em Edifícios e Pontes (RSA), Ministério da Habilitação, Obras Públicas e Transportes, I Serie, No. 125, Lisbon, Portugal, p. 34 (1983)

29. Technical Standards for Constructions in Seismic Areas. Ministry of Public Works and Ministry of the Interior, Ministerial Decree 3 March 1975 (DM 1975). Rome, Italy, p. 12 (1975)

30. B.F. Soysal, B.O. Ay, Y. Arici, Evaluation of the ground motion scaling procedures for concrete gravity dams. Procedia Eng. **199**, 844–849 (2017)

31. E. Zacchei, R. Brasil, Semi-active tuned mass dampers under combined variable actions of friction forces and external disturbances. Arab. J. Geosci. **16**, 1–18 (2023)

32. Norma de la Construcción Sismorresistente - Parte general y edificación (NCSE-02), p. 70. Ministro de Fomento, Madrid, España (2002)

33. A.H. Barbat, L. Orosco, J. E. Hurtado, M. Galindo, *Definition of Seismic Action. Monographic of Seismic Engineering* (International Center for Numerical Methods in engineering, CIMNE IS-10, Barcelona, Spain, 1994), p. 130

34. G. Barone, F. Lo Iacono, G. Navarra, A. Palmeri, A novel analytical model of power spectral density function coherent with earthquake response spectra, in *1st ECCOMAS Thematic Conference on Uncertainty Quantification in Computational Sciences and Engineering*, Crete, Greece, 25–27 May (2015)

35. E. Zacchei, J.L. Molina, Application of artificial accelerograms to estimating damage to dams using failure criteria. Scientia Iranica **27**, 2740–2751 (2020)

36. P.S. Koutsourelakis, A note on the first-passage problem and Vanmarcke's approximation—short communication. Probab. Eng. Mech. **22**, 22–26 (2007)

37. E. Zacchei, J.L. Molina, Damage estimation on concrete gravity dams through artificial accelerograms. MATEC Web Conf. **211**, 1–6 (2018)

38. S.M. Ross, *Probability and Statistics for Engineers and Scientists* (2nd edn.) (Apogeo Editor, Italy, 2008), p. 614

39. M.R. Falamarz-Sheikhabadi, Simplified relations for the application of rotational components to seismic design codes. Eng. Struct. **59**, 141–152 (2014)

40. C.A. Cornell, Engineering seismic risk analysis. Bull. Seismol. Soc. Am. **58**, 1583–1606 (1968)

41. J. Douglas, Ground motion prediction equations 1964–2021, Report. Accessed on Nov 2024. https://rapidn.jrc.ec.europa.eu/reference/152

42. IGME: ZESIS: Base de Datos de Zonas Sismogénicas de la Península Ibérica y territorios de influencia para el cálculo de la peligrosidad sísmica en España (2015). Accessed: June 2023. http://info.igme.es/zesis/

43. A.A.D. De Almeida, M. Assumpção, J.J. Bommer, S. Drouet, C. Riccomini, C.L.M. Prates, Probabilistic seismic hazard analysis for a nuclear power plant site in Southeast Brazil. J. Seismol. **1–23**, 2018 (2018)
44. E. Zacchei, J.L. Molina, Probabilistic seismic hazard analysis for Andalusian Dams in Southern Spain using new seismogenic zones. ASCE-ASME J. Risk Uncertainty Eng. Syst., Part A: Civ. Eng. **8**(3), 1–13 (2022)
45. E. Zacchei, J.L. Molina, R. Brasil, Seismic hazard assessment of arch dams via dynamic modelling: an application to the Rules Dam in Granada, SE Spain. Int. J. Civil Eng. **17**, 323–332 (2019)
46. DISS Working Group, Database of Individual Seismogenic Sources (DISS), version 3.3.0: a compilation of potential sources for earthquakes larger than M 5.5 in Italy and surrounding areas. Istituto Nazionale di Geofisica e Vulcanologia (INGV) (2021). Accessed Oct 2024. https://diss.ingv.it/mapper/
47. Global earthquake model (GEM), database, OpenQuake Map Viewer. Accessed in Nov 2024. https://www.globalquakemodel.org/gem
48. E.M. Scordilis, Empirical global relations converting Ms and mb to moment magnitude. J. Seismol. **10**, 225–236 (2006)
49. J.M. Gaspar-Escribano, A. Rivas-Medina, H. Parra, L. Cabañas, B. Benito, S. Ruiz Barajas, J.M. Martínez Solares, Uncertainty assessment for the seismic hazard map of Spain. Eng. Geol. **199**, 62–73 (2015)
50. International Commission on Large Dams (ICOLD), Selecting Seismic Parameters for Large Dams; Guidelines, Bulletin 148, Paris, France (2016)
51. E. Faccioli, R. Paolucci, *Elementi di sismologia applicata all'ingegneria* (Pitagora Editrice, Bologna, Italy, 2005), p. 255
52. N.N. Ambraseys, K.A. Simpson, J.J. Bommer, Prediction of horizontal response spectra in Europe. Earthq. Eng. Struct. Dyn. **25**, 371–400 (1996)
53. F. Sabetta, A. Pugliese, Estimation of response spectra and simulation of nonstationary earthquake ground motions. Bull. Seismol. Soc. Am. **86**, 337–352 (1996)
54. L. Furgani, Verifiche Sismiche di Dighe in Calcestruzzo, Ph.D. Thesis, p. 350, University of Roma Tre, Rome, Italy (2014)
55. L. Hutchings, G. Viegas, Application of empirical Green's functions in earthquake source, wave propagation and strong ground motion studies. Earthq. Res. Anal. New Front. Seismol. **3**, 87–140 (2012)
56. K. Aki, P.G. Richards, *Quantitative Seismology*, 2nd edn. (Ed. Jane Ellis, 2002), p. 743
57. E. Zacchei, R. Brasil, A new approach for physically based probabilistic seismic hazard analyses for Portugal. Arab. J. Geosci. **15**, 1–22 (2022)
58. E. Fergany, L. Hutchings, Demonstration of pb-PSHA with Ras-Elhekma earthquake. Egypt. NRIAG J. Astr. Geoph. **6**, 41–51 (2017)
59. L. Hutching, A. Mert, Y. Fahjan, T. Novikova, A. Golara, M. Miah, E. Fergany, W. Foxall, Physics-based hazard assessment for critical structures near large earthquake sources. Pure Appl. Geophys. **174**, 3635–3662 (2017)
60. E. Zacchei, J.L. Molina, R. Brasil, Nonlinear degradation analysis of arch-dam blocks by using deterministic and probabilistic seismic input. J. Vib. Eng. Technol. **7**, 301–309 (2019)
61. E. Zacchei, J.L. Molina, R. Brasil, Seismic hazard and structural analysis of the concrete arch dam (Rules Dam on Guadalfeo River). Procedia Eng. **199**, 1332–1337 (2017)
62. U.S. Army Corps of Engineers (USACE), Arch Dam Design, Manual No. 1110-2-2201, Washington, D.C., p. 240 (1994)
63. E. Zacchei, Progettazione definitiva-esecutiva attraverso modelli numerici di una banchina del porto di nuova costruzione "Astilleros del Alba (Astialba)" in Venezuela, p. 163. Thesis of the master's degree, University of Roma Tre, Rome, Italy (2014)
64. E. Zacchei, P. Lyra, F. Stucchi, Inelastic static analysis to estimate the displacement performance for pile-supported wharves. Encontro nacional Betão Estrutural, BE 2018, LNEC Libson, Portugal, 1–9, 7–9 of Nov 2018

65. Metodologia para a avaliação da segurança sísmica de edifícios existentes baseada em analises de fiabilidade estrutural – Edifício de betão armado, Report 81/2019, LNEC, Lisbon, Portugal, Artigo 1° da Portaria n° 302/2019, de 12 de setembro de 2019. http://www.lnec.pt/pt/servicos/ferramentas/

66. Regulamento de Estruturas de Betão Armado e Pré-Esforçado (REBAP), Ministerio de habitaçao, obras publicas e transportes, Lisbon, Portugal, p. 68 (1983)

67. ITCOLD working group, Comportamento delle dighe ai sismi in altri Paesi: Case histories di rilievo (2019). https://www.itcold.it/gruppo-di-lavoro/comportamento-delle-dighe-ai-sismi-in-altri-paesi-case-histories-di-rilievo/

68. G. Buffi, R. Caruana, A. Balsamo, M. Canci, G. Farronato, R. Figini, M. Palmieri, R. Previtali, E. Taffini, E. Zacchei, Seismic behaviour of earth dams: world relevant cases, in *International Symposium on Dams and Earthquakes, Meeting of the EWG*, Athens, 12–13 Sept , pp. 91–102

69. Spanish Association of Dams and Reservoirs (SEPREM). Accessed on March 2025. https://www.seprem.es/ficha.php?idpresa=946&p=37

70. E. Zacchei, Riesgo sísmico y análisis dinámico de presas arco-gravedad. Ph.D. Thesis, Polytechnic School of the University of Salamanca (USAL), Avila, p. 57 (2018)

71. G.W. Housner, The dynamic behavior of water tanks. Bull. Seismol. Soc. Am. **53**, 381–387 (1963)

72. European Committee for Standardizations (CEM), Eurocode 8: Design of structures for earthquake resistance—Part 4: Silos, tanks and pipelines. European Standard EN 1998-4:2006, European Committee for Standardization, Brussels (2006)

73. E. Zacchei, R. Brasil, Strong motion time-history for storage tanks by using seven different earthquake accelerograms. Revista Interdisciplinar de Pesquisa em Engenharia **19**, 1–5 (2017)

74. E. Zacchei, R. Brasil, Total seismic response of circular oil storage tanks: a simplified analysis, in *Proceedings of the XXXVI Iberian Latin-American Congress on Computational Methods in Engineering, CILAMCE*, Rio de Janeiro, Brazil, Nov 22–25, pp. 1–5 (2015)

75. R.C. Hibbeler, *Structural Analysis* (Pearson Education, Inc., Ciudad de Mexico, Mexico, 2012), p. 720

76. E. Zacchei, R. Brasil, An analytical accurate analysis for uniform and damped soil supported on elastic rock. Arab. J. Geosci. **18**, 1–12 (2025)

77. European Committee for Standardizations (CEM), Eurocode 8: Design of structures for earthquake resistance—Part 2: Bridges, EN 1998-2:2005+A2:2011, Brussels, Belgium

78. California Department of Transportation (CALTRANS), Seismic design criteria, version 1.6, p. 160, Sacramento, California (2010)

79. A.N. Perez, The Rules Dam. Revista de Obras Públicas **3441**, 131–152 (2004)

80. G. Galilei, *Dialogos sobre los Sistemas del Mundo (jornada primera)* (Editor Maxtor, Spain, 2010), p. 198

81. E. Zacchei, R. Brasil, T. Taniguchi, Hydrodynamic pressures and mechanical behaviour of unanchored rigid tanks subjected to rocking accelerations. Int. J. Press. Vessels Pip **219**, 1–12 (2026)

82. J.R. Swaisgood, Behavior of embankment dams during earthquake. The Journal of Dam Safety **12**, 1–10 (2014)

GPSR Compliance
The European Union's (EU) General Product Safety Regulation (GPSR) is a set
of rules that requires consumer products to be safe and our obligations to
ensure this.

If you have any concerns about our products, you can contact us on

ProductSafety@springernature.com

In case Publisher is established outside the EU, the EU authorized
representative is:

Springer Nature Customer Service Center GmbH
Europaplatz 3
69115 Heidelberg, Germany